AF693517

Desanka P. Radunović, Ph. D.

WAVELETS
from MATH to PRACTICE

Springer

ACADEMIC MIND

Desanka P. Radunović, Ph. D.
Faculty of mathematics, University of Belgrade

WAVELETS
From MATH to PRACTICE

Reviewers
Miloš Arsenović, Ph. D.
Boško Jovanović, Ph. D.
Branimir Reljin, Ph. D.

(c) 2009
ACADEMIC MIND, Belgrade, Serbia
SPRINGER-VERLAG, Berlin Heidelberg, Germany

Design of cover page
Zorica Marković, Academic Painter

Printed in Serbia by
Planeta print, Belgrade

Circulation
500 copies

ISBN 978-86-7466-345-5
ISBN 978-3-642-00613-5

Library of Congress Control Number: assigned

NOTICE: No part of this publication may be reprodused, stored in a retreival system, or transmitted in any form or by any means, electronic, mechanical, photocopying, recording or otherwise, without the prior written permission of the publishers. All rights reserved by the publishers.

WAVELETS

from MATH to PRACTICE

To my boys

Joca, Boki and Vlada

Preface

Real world phenomena are permanently changing with various speeds of change. Repeating of four seasons in a year accompanied by appropriate changes in nature, alternation of day and night within twenty four hours, heart pulsations, air vibrations that produce sound or stock-market fluctuations are only several examples. Furthermore, since most of these problems express nonlinear effects characterized by fast and short changes, small waves or wavelets are an ideal modeling tool. An oscillatory property and multiresolution nature of wavelets recommends them for use both in signal processing and in solving complex mathematical models of real world phenomena.

As a professor at the School of Mathematics, who teaches computer science students, I feel the need to bridge the gap between the theoretical and practical aspects of wavelets. On the one side, mathematicians need help to implement wavelet theory in solving practical problems. On the other side, engineers and other practitioners need help in understanding how wavelets work in order to be able to create new or modify the existing wavelets according to their needs. This book tries to satisfy both wavelet user groups; to present and explain the mathematical bases of the wavelet theory and to link them with some of the areas where this theory is already being successfully applied. It is self contained and no previous knowledge is assumed.

The introductory chapter gives a short overview of the development of the wavelet concept from its origins at the beginning of the twentieth century until now. Wavelet theory is a natural extension of the Fourier's harmonic analysis. Therefore, we start by presenting the least-square approximation and various forms of the Fourier transform in Chapter 2. Wavelets and the wavelet transform are introduced at the end of this chapter in order to surpass some deficiencies of the Fourier analysis. Multiresolution, as one of the basic wavelet approximation properties, is defined at the beginning of Chapter 3. A dilatation equation, with a scaling function as its solution, and a wavelet equation follow from the mathematical definition of multiresolution. It is further explained how to obtain an orthogonal wavelet basis and a representation of a square integrable function on such basis. The so-called pyramid algorithm is even more efficient than the famous Fast Fourier algorithm (FFT). The theory elaborated in this chapter is demonstrated on several elementary examples that are given at the end of this chapter. Some properties that are very important for the approximation theory, such as the existence and smoothness of a scaling function and the accuracy of the wavelet approximation, are elaborated in

Chapter 4. This analysis shows how to construct wavelets with desired properties.

The last three chapters are mostly application oriented. A brief review of some well known types of wavelets and a few ideas how to construct new wavelets are given in Chapter 5. The principal application area where wavelets are successfully applied nowadays is signal processing. This is because the coherent wavelet theory was initially derived from the analogy between wavelets and filters, in the eighties of the last century. Consequently, Chapter 6 is devoted to filters, as operators applied on discrete signals, and their relations to wavelets. Special attention is paid to orthogonal filters that generate Daubechies family of wavelets. The last chapter (Chapter 7) illustrates a few of the numerous areas where wavelets are being successfully applied. The wavelet theory is rather young (it has existed for less then thirty years) and there are many open questions related to its research and applications.

Finally, some remarks about the notation used in this book are given. Numeration of theorems, lemmas, definitions, examples and formulas are reinitialized in every chapter. Each statement from a different chapter is referred to by the chapter number and the statement number; for example, (3.24) means formula (24) in Chapter 3, and theorem 3.1 means theorem 1 in Chapter 3. If statements from the same chapter are referred to, the chapter number is omitted.

I would like to express my gratitude to Professors B. Reljin, B. Jovanović and M. Arsenović and graduate student Z. Udovičić for their useful comments on this text.

Belgrade, January 2009 D. P. Radunović

Contents

List of Figures

1

Introduction

Sometimes we need an accurate or approximate representation of a quantity in a different form, either the quantity is given by an analytical expression or by the finite set of it's values. The reason may be that we can do calculations easier (calculate some values, differentiate, integrate or something else) or we can get some new information about the quantity by using the new representation. The new representation need to be close to the original and given in a form adequate for a given problem. Mathematically, a form of the new representation depends on a selected projection space, i.e. depends on it's basis and a selected norm.

One of the most frequently used approximation method in practice (signal processing, differential equation modeling) is the least-squares method, where the similarity between functions on an interval I is measured by the inner product $(f, g) = \int_I f(x)\, g(x)\, dx$ in a continuous case. In a discrete case the integral is substituted by a sum over a given set of argument values. Using transformations, we usually measure the similarity of a given function with an entire class of functions depending on one or more parameters (such as the frequency in the Fourier transform) which may change continuously or discretely. The said class of functions is the basis (or frame) of the projection space, and the goal is to select a basis to represent our function so that the representation provides information about the properties of the function which are important to us.

We shall shortly emphasize the most important novelties that wavelets bring to the least-square approximation technique.

- The wavelet representation is given in the space-frequency domain, opposite to the Fourier analysis that gives only a frequency representation. Compact supports of wavelets provide a space, and their oscillatory nature provides a frequency representation of a transformed function. It is clear that such representation is essential for the non-stationary signal processing, which is prevailing in applications.

- The wavelet representation of a function has the multiresolution property, which means that it is given on several resolution scales. Details defined on various refinement levels (fine meshes) are added to the rough approximation determined on a coarse mesh. If we make a good choice of a basis so that it matches the given function well, corrections (details) can be neglected mostly as they will be small. The dimension of the data set that store information about our function is considerably decreased while the most important information is not lost. This is very important for a good compression that saves storage and time. A data compression is fundamental for a development of information and communication technologies, but also for an efficient mathematical modeling of large-scale processes.

The contemporary wavelet theory defines outlines for construction of wavelets and transformations using them. It gives rules that one has to obey to get a wavelet basis with desired properties, meaning that everyone can create a wavelet adequate for his problem. The aim of this book is to help in understanding this rules.

We shall start with a short history of wavelet ideas. Two centuries ago, in 1807, a famous French mathematician Fourier (Jean-Baptiste Joseph) proposed that every 2π-periodic integrable function is the sum of its "Fourier" series

$$(1) \qquad f(x) \sim \frac{a_0}{2} + \sum_{k=1}^{\infty} (a_k \cos kx + b_k \sin kx), \qquad f \in \mathcal{L}_2[-\pi, \pi]$$

for the corresponding values of the a_k and b_k coefficients (more on this in §2.2). What new information about a function $f(x)$ we can get from the representation (1)? It is clear that we can see whether a function $f(x)$ changes fast or slow, because the expression (1) is given through the oscillatory functions with various frequencies k. Indexes k associated to larger modula $|a_k|$ or $|b_k|$ are dominant frequencies; if these indexes are low, a function f is smooth, and if most of them are high a function f changes fast.

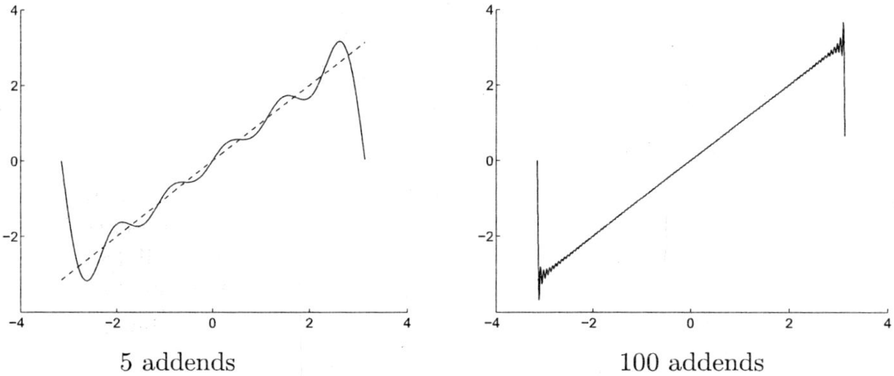

5 addends 100 addends

Figure 1.1: Partial sums of a linear function Fourier series

What shall we do if our function $f(x)$ changes its behavior in time – it is smooth for a while and then it starts to change fast? The representation (1) cannot give us an adequate information in this case, because trigonometric functions $\cos kx$ and $\sin kx$ are not localized in time as they last infinitely. We need basis functions which will be oscillatory, like a sine function, but with finite duration.

Haar [28] wondered whether there exists an orthonormal system of functions on the interval $[0, 1]$ such that, for any function $f(x)$ continuous on that interval, the series

$$(f,\, h_0)\, h_0(x) + (f,\, h_1)\, h_1(x) + \cdots + (f,\, h_n)\, h_n(x) + \cdots$$

(the inner product of functions $(f,\, h)$ is defined by formula (2.1)) uniformly converges towards $f(x)$ on the interval $[0, 1]$? This problem has an infinite number of solutions. Haar, in 1909, provided the simplest one and this event can be concerned as the first beginning of the wavelet theory. For a basis function $h_n(x)$ he chose the characteristic function of the dyadic interval $I_n = [2^{-j}k,\, 2^{-j}(k+1))$, $n = 2^j + k$, which equals one on that interval and equals zero outside of it (Figure 1.2). An approximation of a function $f(x)$ by a partial sum of the said series is nothing else but a well known approximation of a continuous function with a piecewise constant function, where approximation coefficients $(f,\, h_n)$ are the mean values of a function $f(x)$ within the corresponding dyadic intervals. Haar's approximation is applicable for functions which are only continuous, or even only integrable with a square on interval the $[0, 1]$ or, more generally, functions that have a regularity index close to zero.

$$h(x) = \begin{cases} 1, & x \in [0,1) \\ 0, & x \notin [0,1) \end{cases} \qquad \underset{n=2^j+k}{h_n(x) = h(2^j x - k),} \qquad f(x) \sim \sum_n \frac{(f, h_n)}{(h_n, h_n)}\, h_n(x)$$

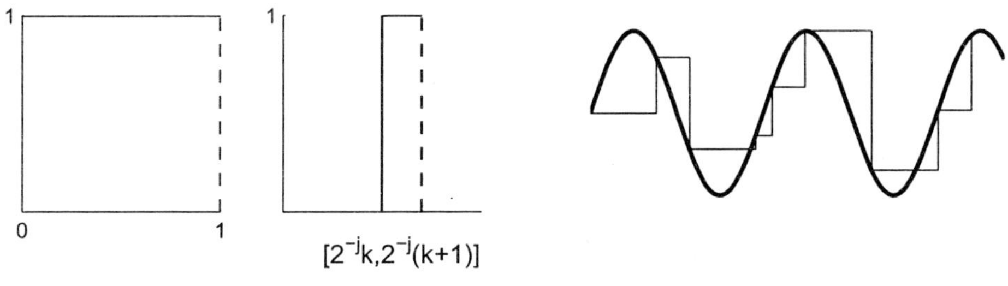

Figure 1.2: Haar decomposition

Ten years later Faber and Schauder (1920) replaced Haar's functions $h_n(x)$ with their primitive functions, the roof functions (Figure 1.3),

$$\Delta_n(x) = \Delta(2^j x - k), \qquad n = 2^j + k, \quad j \geq 0, \quad 0 \leq k < 2^j.$$

where

$$\Delta(x) = \begin{cases} 2x, & x \in [0, 1/2] \\ 2(1-x), & x \in [1/2, 1] \\ 0, & x \notin [0, 1] \end{cases}$$

If we add functions 1 and x we get Shauder's basis

$$1, \quad x, \quad \Delta_1(x), \quad \ldots, \quad \Delta_n(x), \ldots$$

of the space of continuous functions on the interval $[0, 1]$. All continuous functions on that interval may be represented by the series

$$f(x) = a + bx + \sum_{n=1}^{\infty} \alpha_n \Delta_n(x),$$

where $a = f(0)$, $b = f(1) - f(0)$ (because $\Delta_n(0) = \Delta_n(1) = 0$ for $n > 0$), and the coefficients α_n are determined by the values of the function in the dyadic points

$$\alpha_n = f\big((k + 1/2)2^{-j}\big) - \frac{1}{2}\left(f(k2^{-j}) + f\big((k+1)2^{-j}\big)\right), \quad n = 2^j + k.$$

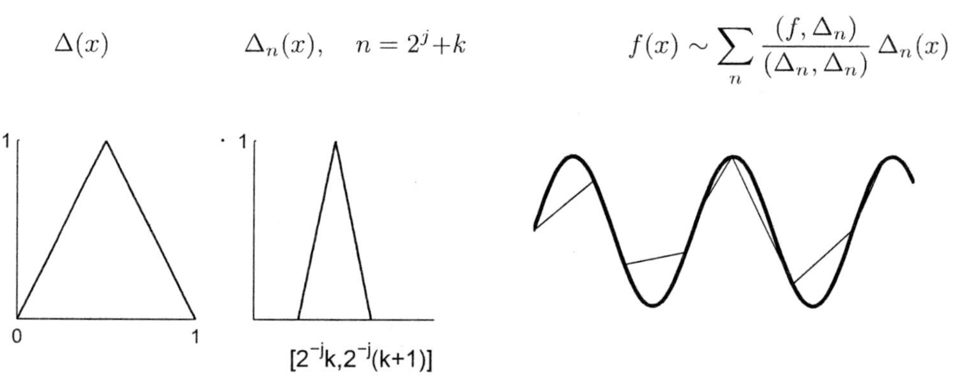

Figure 1.3: Schauder's decomposition

Using Shauder's basis, Paul Levy analyzed the multifractal structure of Brownian motion and obtained results in studying properties of local regularity that were better than those arrived at using Fourier's basis. Schauder's basis implements the idea of multiresolution analysis through the mapping $x \to 2^j x - k$.

Since 1930, only individual contributions that were not part of a coherent theory appeared in the next more than fifty years. The term wavelet and the corresponding theory were not well known, so many specific techniques were subsequently rediscovered by physicists and mathematicians working with wavelets. By applying wavelets to the signal processing in the early eighties of the last century, a coherent theory of wavelets emerged along with their extensive use in various areas. Today, the borders between the mathematical approach and the approach from the perspective of signal and image processing are disappearing. That very connection brought about an enormous advance in this area, as well as wavelets of Ingrid Daubechies, as new special functions.

The name wavelet with its contemporary meaning was first used by Grossman, a physicists and Morlet, an engineer [27] during the early eighties of the last century. Based on physical intuition they defined wavelets within the context of quantum physics. Working on digital signal processing Stephane Mallat [29] provided a new contribution to wavelet theory by connecting the term filters with mirror symmetry (mirror filters), the pyramid algorithm and orthonormal wavelet basis. Yves Meyer [31] constructed a continuously differentiable wavelet lacking in that it does not have a compact support (a finite domain where it is not equal to zero). Finally, Ingrid Daubechies [19] managed to add to Haar's work by constructing various families of orthonormal wavelet bases.

For every integer r Daubechies constructed an orthonormal basis of the space of functions integrable with a square

$$2^{j/2}\psi_r(2^j x - k), \qquad j,\, k \in \mathcal{Z}.$$

It is determined by the function $\psi_r(x)$ with the following properties:

- The compact support of the function $\psi_r(x)$ is the interval $[0, 2r-1]$.

- The function $\psi_r(x)$ has first r moments equal to zero,

$$\int_{-\infty}^{\infty} \psi_r(x)\, dx = \cdots = \int_{-\infty}^{\infty} x^{r-1}\psi_r(x)\, dx = 0.$$

- The function $\psi_r(x)$ has γr continuous derivatives, where $\gamma \approx 0.2$.

Haar's system of functions is Daubechies wavelet family for $r = 1$. Daubechies wavelets provide for a far more efficient analysis or synthesis of a smooth function for a greater r. Namely, if the function being analyzed has m continuous derivatives, where $0 \leq m \leq r$, the coefficients $b_{j,k}$ in the decomposition by Daubechies basis are of the order $2^{-(m+1/2)j}$, and if $m > r$, the coefficients $b_{j,k}$ are of the order $2^{-(r+1/2)j}$. This means that for a regular function the coefficient values for a greater r are much smaller than in case of, for instance, using Haar's system, where these coefficients are of the order $2^{-3j/2}$. This property is essential for data compression, consisting of neglecting small coefficients (small according to a predetermined threshold). It provides a minimum set of remaining coefficients for memorizing data or functions using. The property is local, for Daubechies wavelets have a compact support. Synthesis using Daubechies wavelets of a higher order also provides better results than synthesis with Haar's system, because using Haar's system a smooth function is approximated by a function with finite jumps. Finally, but not less important, decomposition and reconstruction algorithms are fast and efficient due to the orthogonality of Daubechies bases. More about Daubechies wavelets (and filters) in §6.4.

It is important to note that, unlike the Fourier analysis which is based on one set of functions (sine functions) wavelet representation is possible on an infinite many different bases. Wavelet families differ one from another according to the compactness of spatially localized basis functions and their smoothness. The optimum choice of a basis, or a representation, depends on properties we want to analyze in the problem being examined. The chosen basis gives information about a function or a signal which is important in a defined sense.

2

Least-squares approximation

Within this book we shall be dealing with various representations of functions in the space $\mathcal{L}_2$. Hilbert's space $\mathcal{L}_2(p;\, a, b)$ is the space of functions integrable with a square on the interval $[a, b]$,

$$\mathcal{L}_2(p;\, a, b) = \left\{ f \,\middle|\, \int_a^b p(x)\bigl|f(x)\bigr|^2\, dx < \infty \right\}.$$

The function $p(x)$ is called *a weight function.* It is defined on the interval $[a, b]$ and it satisfies the condition $p(x) > 0$ almost everywhere, it can be equal to zero only on a set with a measure of zero. The sign $\mathcal{L}_2(a, b)$ shall be used when the weight function $p(x) \equiv 1$. The number

$$\|f\| = \left(\int_a^b p(x)\bigl|f(x)\bigr|^2\, dx \right)^{1/2}$$

is often called the *energy norm* of the function $f(x)$. Therefore, we can say that $\mathcal{L}_2$ is the space of functions with a finite energy. This norm is induced by the inner product

$$(f\,,\, g) = \int_a^b p(x) f(x) \overline{g(x)}\, dx, \qquad \|f\|^2 = (f, f), \tag{1}$$

where $\overline{g(x)}$ represents the conjugated-complex function of the function $g(x)$.

2.1 Basic notations and properties

The best least-squares approximation for a function $f \in \mathcal{L}_2(p;\, a, b)$ in a subspace $\mathcal{H} \subset \mathcal{L}_2(p;\, a, b)$, defined by linearly independent functions $g_k(x) \in \mathcal{L}_2(p;\, a, b)$, $k = 0, \ldots, n$, is a generalized polynomial

$$Q_n^\circ(x) = c_0^\circ\, g_0(x) + \cdots + c_n^\circ\, g_n(x), \tag{2}$$

that varies the least from the function $f(x)$ regarding the energy norm

$$\|f - Q_n^\circ\| = \inf_{Q_n \in \mathcal{H}} \|f - Q_n\| = \inf_{Q_n \in \mathcal{H}} \left(\int_a^b p\,|f - Q_n|^2\, dx \right)^{1/2}.$$

Therefore, $Q_n^\circ(x)$ is the function from the set of allowed functions

$$Q_n(x) = \sum_{k=0}^{n} c_k g_k(x)$$

whereby the minimum mean variance is achieved. It means that the surface formed by the functions $f(x)$ and $Q_n^\circ(x)$ and the lines $x = a$ and $x = b$ has minimal size, although the variance of the function $Q_n^\circ(x)$ from $f(x)$ may be great in certain points of the interval. Using the function $p(x)$ a varying quality of the approximation is achieved in different parts of the interval. Namely, in parts of the interval where $p(x)$ is greater difference $f(x) - Q_n^\circ(x)$ is multiplied by a greater factor, thus these segments take part in the minimization with a greater weight. This is the reason why the function $p(x)$ is called a weight function.

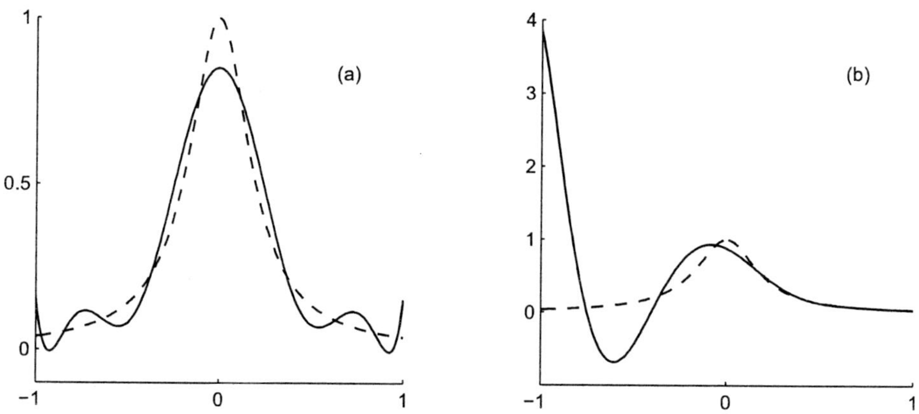

Figure 2.1: Least-squares approximations for different weight functions

EXAMPLE 1. Figure 2.1 shows a ninth degree polynomials of the best least-squares approximation (full line) of the function $f(x) = 1/(1 + 25x^2)$ (dashed line) for the weight functions $p(x) \equiv 1$ (a) and $p(x) = e^{10x}$ (b). ∎

For a function $f(x)$, given only by it's values $f(x_k)$ on the finite set of points x_k, $k = 0, \ldots, m$, the distance cannot be measured using the inner product (1). The integral is substituted by the sum that defines another inner product in the space $\mathcal{L}_2$,

$$(f\,,\,g)=\sum_{k=0}^{m} p_k f(x_k)\,\overline{g(x_k)},\qquad p_k>0. \tag{3}$$

p_k are given positive numbers, called weight coefficients. They have the same role as the weight function $p(x)$ in the continuous case. The inner product (3) defines the *discrete energy norm*

$$\|f\|^2=(f\,,\,f)=\sum_{k=0}^{m} p_k\big|f(x_k)\big|^2. \tag{4}$$

of the Hilbert space $\mathcal{L}_2$. The best least-squares approximation $Q_n^\circ(x)$ varies the least from the function $f(x)$ regarding the discrete energy norm

$$\|f-Q_n^\circ\|=\inf_{Q_n\in\mathcal{H}}\|f-Q_n\|=\inf_{Q_n\in\mathcal{H}}\left(\sum_{k=0}^{m} p_k\big|f(x_k)-Q_n(x_k)\big|^2\right)^{1/2},\qquad m>n.$$

Irrespective of the used norm the best approximation $Q_n^\circ(x)$ always exists and is uniquely defined because every Hilbert space is a strictly normed linear space,

$$\|f+g\|=\|f\|+\|g\|\qquad\Longleftrightarrow\qquad g=\lambda f,\quad \lambda\in\mathcal{R}.$$

We should note that standard symbols are used in this book: $\mathcal{Z}$ is the set of integer, $\mathcal{R}$ the set of real and $\mathcal{C}$ the set of complex numbers.

LEMMA 1. *$Q_n^\circ(x)$ is the best approximation of a function $f(x)\in\mathcal{L}_2(p;\,a,b)$ in a subspace $\mathcal{H}$ if and only if $(f-Q_n^\circ,\,Q_n)=0$ for every function $Q_n\in\mathcal{H}$.*

Proofs for these statements can be found in [37].

Lemma 1 claims that $Q_n^\circ(x)$ represents an orthogonal projection of the function $f(x)$ to the subspace $\mathcal{H}$. Thus, an arbitrary function $Q_n(x)$ may be replaced with basis functions $g_j(x)$, $j=0,\dots,n$, of the subspace $\mathcal{H}$ in the orthogonality condition,

$$(f-Q_n^\circ\,,\,g_j)=0,\quad j=0,\dots,n.$$

It follows that coefficients in the representation (2) are the solution of the system of linear equations

$$\sum_{k=0}^{n} c_k^\circ\,(g_k,\,g_j)=(f,\,g_j)\qquad j=0,\dots,n. \tag{5}$$

The determinant of the system matrix is the *Gramm determinant*

$$G(g_1,\dots,g_n)=\det\big((g_k\,,\,g_j)\big),$$

and is different from zero because we assumed that functions $g_k(x)$, $k = 0, \ldots, n$, are linearly independent.

Since the system (5) is more ill conditioned as the dimension of the system increases, it is preferable to use orthonormal function systems. The basis $\{g_k\}_{k=0}^n$ of a finite-dimensional space is called an *orthonormal basis* if basis elements meet conditions (Figure 2.2(a))

$$(g_k\,,\, g_j) = \delta(k-j) = \begin{cases} 1, & k = j, \\ 0, & k \neq j, \end{cases} \qquad k, j = 0, \ldots, n.$$

In this case the matrix of the system (5) is the identity matrix and the solution of the system are *Fourier coefficients* of the function $f(x)$ according to the orthonormal function system $\{g_k(x)\}_{k=0}^n$,

$$c_k^{\circ} = (f,\, g_k)\,, \qquad k = 0, \ldots, n. \tag{6}$$

The best approximation according to the orthonormal basis is then given by the expression

$$Q_n^{\circ}(x) = \sum_{k=0}^{n} (f,\, g_k)\, g_k(x).$$

When $n = \infty$ and the countable orthonormal system of functions $\{g_k(x)\}_{k=0}^{\infty}$ is complete, the function $f(x)$ is represented by its *Fourier series*

$$f(x) = \sum_{k=0}^{\infty} (f,\, g_k)\, g_k(x). \tag{7}$$

The countable orthonormal system of elements is complete if there is no other element of the space, different from zero, which is orthogonal to all elements of the system. Series (7) converges to the function $f(x)$ in the $\mathcal{L}_2$ norm according to the following lemma [37],

LEMMA 2. *In a Hilbert space the Fourier series of an arbitrary element per a complete orthonormal system of elements converges to that element.*

The Parseval equality expresses the equality of the energy norms of a function $f(x)$ and the vector of its Fourier coefficients (6),

$$\|f\|^2 = \sum_{k=0}^{\infty} |(f,\, g_k)|^2, \tag{8}$$

and it is a consequence of lemma 2.

The Generalized Parseval equality is expressed by the inner product

$$(f,\, h) = \sum_{k=0}^{\infty} \overline{(f,\, g_k)}\, (h,\, g_k).$$

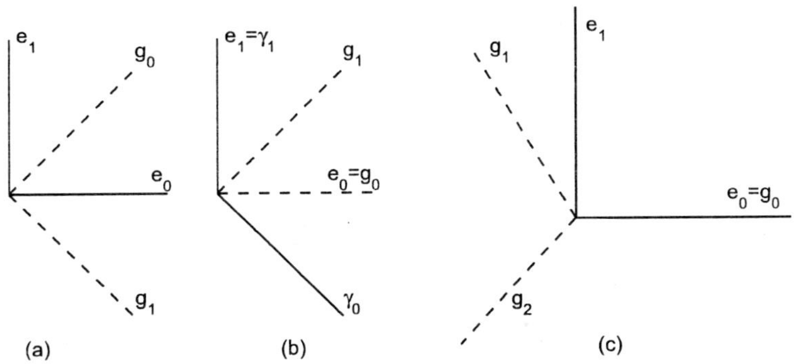

Figure 2.2: Bases in $\mathcal{R}^2$

In the text we shall also use the following types of bases:

Biorthogonal bases are two full sets of linearly independent elements $\{g_k\}$ and $\{\gamma_k\}$ of a Hilbert space such that (Figure 2.2(b))

$$(g_k,\, \gamma_j) = \delta(k-j) \tag{9}$$

The Parseval equality for biorthogonal bases has the form of

$$\|f\|^2 = \sum_k \overline{(f,\, g_k)}\,(f,\, \gamma_k),$$

and the generalized Parseval equality is equal to

$$(f,h) = \sum_k \overline{(f,\, g_k)}\,(h,\, \gamma_k) = \sum_k \overline{(f,\, \gamma_k)}\,(h,\, g_k).$$

A Riesz basis (stable basis) is a countable set of elements $\{g_k\}$ of a Hilbert space that meet the condition that all elements f of this space may be uniquely represented as a sum $f = \sum_k c_k g_k(x)$, where there are positive constants A and B such that

$$A\|f\|^2 \le \sum_k |c_k|^2 \le B\|f\|^2, \quad 0 < A \le B < \infty. \tag{10}$$

In a finitely dimensional space all bases are Riesz bases. An orthonormal basis is a Riesz basis with the constants $A = B = 1$, according to (8). The basis $1,\, x,\, x^2, \ldots$ is not a Riesz basis in $\mathcal{L}_2(0,1)$ because the constant $A = 0$. The inner products $(x^k\,,\, x^l) = 1/(k+l+1)$ are elements of an ill conditioned Hilbert matrix, thus the infinitely dimensional Hilbert matrix is not positively definite.

A frame is a complete, but predefined set of elements $\{g_k\}$ of the Hilbert space (the elements are linearly dependent, Figure 2.2(c)), and

$$A\|f\|^2 \le \sum_k |(f,\, g_k)|^2 \le B\|f\|^2, \quad 0 < A, B < \infty$$

The frame is *tight* if the following condition is met

$$\sum_k |(f, g_k)|^2 = A\|f\|^2, \qquad \text{and so} \qquad f(x) = A^{-1} \sum_k (f, g_k)\, g_k(x).$$

This representation is not unique, because the expression $\sum_k \beta_k g_k(x) = 0$ may be added to it, which is a consequence of the linear dependence of the frame elements.

Let us return to the Fourier series (7) of a function $f(x)$. For certain forms of the weight function $p(x)$ the orthonormal function systems are known [1]. For example,

- The system of Legendre polynomials is orthogonal on the interval [-1, 1] in relation to the weight function $p(x) \equiv 1$.
- The system of Chebyshev polynomials of the first kind is orthogonal on the interval $[-1, 1]$ in relation to the weight function $p(x) = 1/\sqrt{1-x^2}$.
- The system of Hermite polynomials is orthogonal on the interval $[-\infty, \infty]$ in relation to the weight function $p(x) = e^{-x^2}$.
- The system of trigonometric functions is orthogonal on the interval $[-\pi, \pi]$ in relation to the weight function $p(x) \equiv 1$,.

2.2 Fourier analysis

It has been noted in the introduction that Fourier discovered that other functions may be represented through the superposition of sines and cosines,

$$f(x) = \frac{a_0}{2} + \sum_{k=1}^{\infty} (a_k \cos kx + b_k \sin kx), \tag{11}$$

i.e. the superposition of the harmonics of various frequencies (Figure 2.3).

Due to the orthogonality of the system of functions

$$1,\ \sin x,\ \cos x,\ \sin 2x,\ \cos 2x, \ldots,\ \sin nx,\ \cos nx, \ldots,$$

the matrix of the system (5), that defines the coefficients of the decomposition (11), is diagonal and thus these coefficients are determined using the formulae

$$\begin{aligned} a_k &= \frac{(f\,,\, \cos kx)}{(\cos kx\,,\, \cos kx)} = \frac{1}{\pi}\int_{-\pi}^{\pi} f(x)\cos kx\, dx \qquad k = 0, \ldots, n, \\ b_k &= \frac{(f\,,\, \sin kx)}{(\sin kx\,,\, \sin kx)} = \frac{1}{\pi}\int_{-\pi}^{\pi} f(x)\sin kx\, dx \qquad k = 1, \ldots, n. \end{aligned} \tag{12}$$

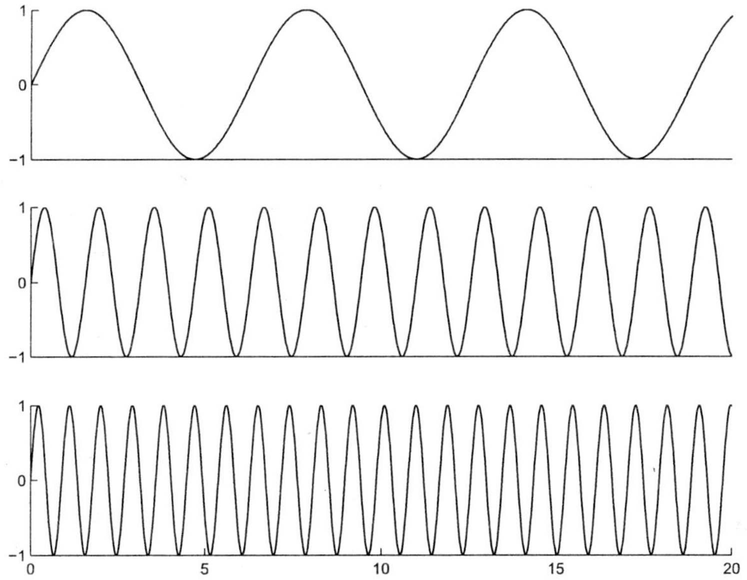

Figure 2.3: Components in the Fourier representation

Therefore, every sufficiently smooth periodic function can be represented by its trigonometric Fourier series (11), i.e. it can be displayed as a linear combination of the sine functions $\sin kx$ and $\cos kx$, $k = 1, 2, \ldots,$ with a frequency of oscillation equal to k, on an interval of 2π. The constant term $\frac{a_0}{2}$ is the mean value of the function $f(x)$ on the interval $(-\pi, \pi)$,

$$f_{mean} = \frac{a_0}{2} = \frac{1}{2\pi} \int_{-\pi}^{\pi} f(x)\, dx,$$

and the other addends in series (11) oscillate around zero and their sum is equal to $f - f_{mean}$. Representation of a function in a frequency domain is called *Fourier* or *harmonic analysis.*

By replacing the functions $\sin kx$ and $\cos kx$ in series (11) with functions of a complex variable

$$\sin kx = \frac{1}{2\imath}\left(e^{\imath kx} - e^{-\imath kx}\right), \qquad \cos kx = \frac{1}{2}\left(e^{\imath kx} + e^{-\imath kx}\right),$$

($\imath = \sqrt{-1}$ is the imaginary unit) we get the Fourier series written in the complex form

$$f(x) = \sum_{k=-\infty}^{\infty} c_k e^{\imath kx}. \tag{13}$$

The system of functions $\{e^{\imath kx}\}_k$ is a complete orthogonal system of functions on the interval $[-\pi, \pi]$,

$$(e^{\imath kx}, e^{\imath lx}) = \int_{-\pi}^{\pi} e^{\imath kx} e^{-\imath lx}\, dx = \begin{cases} 0, & \text{za } k \neq l, \\ 2\pi, & \text{za } k = l. \end{cases}$$

Thus the Fourier coefficients in representation (13) are equal to

$$(14) \qquad c_k = \frac{(f\,,\, e^{\imath kx})}{(e^{\imath kx}\,,\, e^{\imath kx})} = \frac{1}{2\pi} \int_{-\pi}^{\pi} f(x) e^{-\imath kx}\, dx, \qquad k = 0, \pm 1, \ldots,$$

and series (13), based on lemma 2, converges in the $\mathcal{L}_2$ norm to the function $f(x)$ it is attached to. The series of coefficients $\{c_k\}$ represents the *spectrum* of the function $f(x)$, and the Fourier analysis is often referred to as the spectral analysis. According to the Parseval equality (8) energy norms of the function and its spectrum are equal

$$\|f\|^2 = \sum_{k=-\infty}^{\infty} |c_k|^2.$$

Other than in the spectral analysis, the representation (13) is also useful in other applications due to positive properties of the functions $e^{\imath kx}$. Namely, these functions are the eigenfunctions of the differential operator and the finite differences operator,

$$\frac{d}{dx} e^{\imath kx} = \imath k\, e^{\imath kx}, \qquad \Delta\, e^{\imath kx} = \left(\frac{e^{\imath kh} - 1}{h}\right) e^{\imath kx}.$$

Therefore, using the representation (13) the problem formulated by a differential or a difference equation can be reduced to a problem formulated by an algebraic equation.

The series (13), with the coefficients given by the expression (14), is attached to a 2π-periodic function $f(x)$. In order to get the appropriate representation for a function periodic on an interval with a length of T, we introduce the substitution $x = 2\pi\, t/T$ into the formula (14)

$$c_k = \frac{1}{T} \int_{-\frac{T}{2}}^{\frac{T}{2}} f\Big(\frac{2\pi}{T}\, t\Big) e^{-\imath k \frac{2\pi}{T}\, t}\, dt = \frac{1}{T} \int_{-\frac{T}{2}}^{\frac{T}{2}} f_T(t) e^{-\imath \omega t}\, dt,$$

where $f_T(t) \equiv f(2\pi\, t/T)$ is a periodic function with the period of T and $\omega = 2\pi\, k/T$. By introducing said substitutions and the symbol $\Delta\omega = 2\pi/T$ into the expression (13), we obtain the Fourier series of the function $f_T(t)$,

$$(15) \qquad f_T(t) = \sum_{k=-\infty}^{\infty} c_k e^{\imath \omega t} = \sum_{k=-\infty}^{\infty} \frac{\Delta\omega}{2\pi} e^{\imath \omega t} \left(\int_{-\frac{T}{2}}^{\frac{T}{2}} f_T(\tau) e^{-\imath \omega \tau}\, d\tau \right).$$

When $T \to \infty$ the function $f_T(t)$ tends to the non-periodic function $F(t) = \lim_{T\to\infty} f_T(t)$ and the sum in the expression (15) tends to the integral by ω, because

$$\frac{1}{T} = \frac{\Delta\omega}{2\pi} \to 0 \qquad \text{when } T \to \infty.$$

The limiting form of the expression (15) is equal to

$$F(t) = \int_{-\infty}^{\infty} \frac{d\omega}{2\pi} e^{\imath\omega t} \left(\int_{-\infty}^{\infty} F(\tau) e^{-\imath\omega\tau}\, d\tau \right). \tag{16}$$

The term in brackets in the formula (16) is called the *Fourier transform* of the function $F(x)$ and is a function of the frequency ω,

$$\hat{F}(\omega) = \int_{-\infty}^{\infty} F(x) e^{-\imath\omega x}\, dx. \tag{17}$$

The expression (16), with (17) taken into consideration, is the *inverse Fourier transform* whereby the function $\hat{F}(\omega)$ is transformed back into the function $F(x)$,

$$F(x) = \frac{1}{2\pi} \int_{-\infty}^{\infty} \hat{F}(\omega) e^{\imath\omega x}\, d\omega. \tag{18}$$

The Parseval equality (8), dealing with maintaining the energy norm while performing a Fourier transform, is valid in this limiting case as well. In order to prove it, let us define the term *convolution of functions.*

DEFINITION 1. *The convolution $f * g$ of functions f and g is a function of the variable x defined by the integral*

$$(f * g)(x) = \int_{-\infty}^{\infty} f(t)\, g(x-t)\, dt. \tag{19}$$

EXAMPLE 2. The convolution of the characteristic function $\aleph_{(0,1)}(x)$ of the interval $(0,1)$ and a continuous function $f(x)$ is the mean value of the continuous function on the interval $(x-1, x)$. Indeed, since the function $\aleph_{(0,1)}(x) = 1$ only when $0 \le x < 1$, and for all other argument values it equals zero, then

$$(\aleph_{(0,1)} * f)(x) = \int_{-\infty}^{\infty} \aleph_{(0,1)}(t) f(x-t)\, dt = \int_0^1 f(x-t)\, dt = \int_{x-1}^{x} f(t)\, dt.$$

∎

For $g(x) = e^{\imath\omega x}$ convolution is, in accordance with (17), equal to

$$(f * g)(x) = \int_{-\infty}^{\infty} f(t) e^{\imath\omega(x-t)}\, dt = e^{\imath\omega x} \int_{-\infty}^{\infty} f(t) e^{-\imath\omega t}\, dt = \hat{f}(\omega)\, e^{\imath\omega x}.$$

This means that complex exponential functions $e^{\imath\omega x}$ are the eigenfunctions of the convolution operator, which adds to the list of nice properties of these functions. The corresponding eigenvalue is the Fourier transform $\hat{f}(\omega)$ for a given frequency ω.

THEOREM 1. (CONVOLUTION THEOREM) *The Fourier transform of the convolution of two functions is equal to the product of their Fourier transforms,*

$$(20)\qquad \widehat{(f * g)}(\omega) = \hat{f}(\omega)\,\hat{g}(\omega).$$

Proof: The statement follows based on definitions (17) and (19),

$$\begin{aligned}\widehat{(f * g)}(\omega) &= \int_{-\infty}^{\infty} (f * g)(x) e^{-\imath\omega x}\,dx = \int_{-\infty}^{\infty} \left(\int_{-\infty}^{\infty} f(t) g(x-t)\,dt\right) e^{-\imath\omega x}\,dx \\ &= \int_{-\infty}^{\infty} f(t) \left(\int_{-\infty}^{\infty} g(x-t) e^{-\imath\omega x}\,dx\right) dt \\ &= \int_{-\infty}^{\infty} f(t) \left(\int_{-\infty}^{\infty} g(u) e^{-\imath\omega(u+t)}\,du\right) dt \\ &= \left(\int_{-\infty}^{\infty} f(t) e^{-\imath\omega t}\,dt\right) \left(\int_{-\infty}^{\infty} g(u) e^{-\imath\omega u}\,du\right) = \hat{f}(\omega)\,\hat{g}(\omega)\end{aligned}$$

∎

Unlike the convolution theorem, the *modulation theorem* expresses the Fourier transform of the product of two functions as the convolution of their Fourier transforms,

$$\widehat{(f\,g)}(\omega) = \frac{1}{2\pi}\,(\hat{f} * \hat{g})(\omega).$$

Modulation is translation in the frequency domain.

Let us now prove the Parseval equality of the Fourier transform.

THEOREM 2. *The Parseval equality for a function $f(x)$ and its Fourier transform $\hat{f}(\omega)$ is*

$$(21)\qquad \int_{-\infty}^{\infty} |f(x)|^2\,dx = \frac{1}{2\pi} \int_{-\infty}^{\infty} |\hat{f}(\omega)|^2\,d\omega.$$

Proof: The Fourier transform of the function $g(x) = \overline{f(-x)}$ equals

$$\hat{g}(\omega) = \int_{-\infty}^{\infty} \overline{f(-x)} e^{-\imath\omega x}\,dx = \int_{-\infty}^{\infty} \overline{f(x)} e^{\imath\omega x}\,dx = \overline{\int_{-\infty}^{\infty} f(x) e^{-\imath\omega x}\,dx} = \overline{\hat{f}(\omega)},$$

and convolution (19) in the point $x = 0$, for the said choice of the function $g(x)$, is

$$(22)\qquad (f * g)(0) = \int_{-\infty}^{\infty} f(t)\,g(-t)\,dt = \int_{-\infty}^{\infty} f(t)\,\overline{f(t)}\,dt = \|f\|^2.$$

On the other hand, the inverse Fourier transform (18) of the equality (20) is given by the expression

$$(f * g)(x) = \frac{1}{2\pi} \int_{-\infty}^{\infty} \hat{f}(\omega)\,\hat{g}(\omega) e^{\imath\omega x}\,d\omega,$$

providing another expression for (22) at the point $x = 0$,

$$(f * g)(0) = \frac{1}{2\pi}\int_{-\infty}^{\infty} \hat{f}(\omega)\,\hat{g}(\omega)e^{0}\,d\omega = \frac{1}{2\pi}\int_{-\infty}^{\infty} \hat{f}(\omega)\,\overline{\hat{f}(\omega)}\,d\omega = \frac{1}{2\pi}\|\hat{f}\|^2.$$

Equating the last expression with (22) we obtain the Parseval equality (21). ∎

By use of the Parseval equality (21) the derivative of a function can be expressed by its Fourier transform. Function $f(x)$ has a derivative of the order s in $\mathcal{L}_2$ if

$$\|f^{(s)}\|^2 = \frac{1}{2\pi}\int_{-\infty}^{\infty} |\omega|^{2s}|\hat{f}(\omega)|^2\,d\omega < \infty.$$

Differentiation in the frequency domain boils down to multiplication by the factor $\imath\omega$. This definition enables s to be a fraction (and a negative one).

The generalization of the Parseval equality (21) for the Fourier transform expresses the equality of inner products in the temporal and frequency domain [49],

$$(f,\,g) = \frac{1}{2\pi}(\hat{f},\,\hat{g}), \quad \text{i.e.} \quad \int_{-\infty}^{\infty} f(x)\,\overline{g(x)}\,dx = \frac{1}{2\pi}\int_{-\infty}^{\infty} \hat{f}(\omega)\,\overline{\hat{g}(\omega)}\,d\omega. \tag{23}$$

In accordance with the terms introduced for the Fourier (17) and inverse Fourier transform (18), the Fourier series (13) can be seen as a discrete variant of the inverse Fourier transform: the frequency is discrete, $\omega = k$, thus $\hat{f}(\omega) \sim c_k$.

In practice, the variable x is often discrete. Namely, a function $f(x)$ is not given for every x, but only for discrete values of the independent variable e.g. $x = n$, in the form of a sequence $f(n)$. If x is a discrete variable, we are dealing with a *Discrete time Fourier transform* (ω is continuous) and a *Discrete time Fourier series* (ω is discrete as well).

The term *Discrete Fourier Transform (DFT)* denotes a transformation of a single period of a periodic function with a discrete argument. Our aim is to determine the N-dimensional vector of the Fourier coefficients $\hat{\mathbf{f}} = \{\hat{f}(k)\}_{k=0}^{N-1}$ for the known vector $\mathbf{f} = \{f(n)\}_{n=0}^{N-1}$ of the function values, or vice versa. It means that we interpolate function $f(x)$ on the set of nodes $n = 0, 1, \ldots, N-1$, by the partial sum of its Fourier series,

$$f(n) = \frac{1}{N}\sum_{k=0}^{N-1} \hat{f}(k)\,e^{\imath k n \frac{2\pi}{N}}, \qquad n = 0, \ldots, N-1. \tag{24}$$

If we denote the N-th root of 1 in the complex plane with

$$W = e^{\imath\frac{2\pi}{N}} = \sqrt[N]{e^{\imath\,2\pi}}, \tag{25}$$

the unknown vector $\hat{\mathbf{f}}$ is the solution of the N-dimensional system of linear equations

$$\mathbf{f} = \frac{1}{N}\mathcal{F}_N\,\hat{\mathbf{f}}, \tag{26}$$

with the *Fourier matrix*

$$\mathcal{F}_N = \begin{pmatrix} 1 & 1 & 1 & \dots & 1 \\ 1 & W & W^2 & \dots & W^{N-1} \\ 1 & W^2 & W^4 & \dots & W^{2(N-1)} \\ \vdots & \vdots & \vdots & \ddots & \vdots \\ 1 & W^{N-1} & W^{2(N-1)} & \dots & W^{(N-1)^2} \end{pmatrix}, \tag{27}$$

As the Fourier matrix is a unitary matrix (see [37] for the proof),

$$\mathcal{F}_N \, \mathcal{F}_N^* = \mathcal{F}_N^* \, \mathcal{F}_N = N \, I, \tag{28}$$

where I denotes the identity matrix, the solution of the system (26) is directly obtained

$$\hat{\mathbf{f}} = \mathcal{F}_N^* \, \mathbf{f},$$

i.e.

$$\hat{f}(k) = \sum_{n=0}^{N-1} f(n) \, W_N^{-nk} = \sum_{n=0}^{N-1} f(n) \, e^{-\imath nk \frac{2\pi}{N}}, \qquad k = 0, \dots, N-1. \tag{29}$$

It is obvious that the *Inverse Discrete Fourier Transform (IDFT)* is given by the expression (24), i.e. (26).

In an analogy to the continuous case (definition 1), we shall define the convolution of two functions of a discrete argument. We shall call such functions discrete *signals*, or just signals. A discrete signal is a number sequence, infinite, periodic or finite, which represents some physical variable changing in time, space, or by some other independent variable.

DEFINITION 2. *The discrete convolution* $\mathbf{h} * \mathbf{x}$ *of signals* $\mathbf{h} = \{h(n)\}$ *and* $\mathbf{x} = \{x(n)\}$ *is the signal* $\mathbf{y} = \{y(n)\}$

$$\mathbf{y} = \mathbf{h} * \mathbf{x}, \qquad y(n) = \sum_k h(k) \, x(n-k). \tag{30}$$

If we suppose that $\mathbf{h} = \{h(n), \, n \geq 0\}$, the convolution of two signals can be calculated by multiplying a signal $\mathbf{x}$ with the lower-triangular matrix generated by the signal $\mathbf{h}$, called the filter matrix in signal theory

$$F = \begin{pmatrix} \cdot & \cdot & \cdot & \cdot & \cdot & \cdot \\ \cdot & h(0) & 0 & 0 & 0 & \cdot \\ \cdot & h(1) & h(0) & 0 & 0 & \cdot \\ \cdot & h(2) & h(1) & h(0) & 0 & \cdot \\ \cdot & h(3) & h(2) & h(1) & h(0) & \cdot \\ \cdot & \cdot & \cdot & \cdot & \cdot & \cdot \end{pmatrix}, \tag{31}$$

so that

$$
\begin{aligned}
&\vdots \\
y(0) &= \cdots + h(N)\,x(-N) + \cdots + h(1)\,x(-1) + h(0)\,x(0) + \ldots \\
&\vdots \\
y(n) &= h(N)\,x(n-N) + \cdots + h(1)\,x(n-1) + h(0)\,x(n) + \ldots \\
&\vdots
\end{aligned}
$$

If $\mathbf{h}$ is a periodic signal with a period N, the matrix F is a $N \times N$-dimension cyclic (Toeplitz) matrix,

$$
F = \begin{pmatrix}
h(0) & h(N-1) & h(N-2) & \cdots & h(1) \\
h(1) & h(0) & h(N-1) & \cdots & h(2) \\
h(2) & h(1) & h(0) & \cdots & h(3) \\
\vdots & \vdots & \vdots & \ddots & \vdots \\
h(N-1) & h(N-2) & h(N-3) & \cdots & h(0)
\end{pmatrix}. \tag{32}
$$

The convolution theorem (Theorem 1.) is true in the discrete case, too.

THEOREM 3. (DISCRETE CONVOLUTION THEOREM) *The Fourier transform of the discrete convolution of two signals equals to the product of Fourier transforms of these signals,*

$$
\widehat{(\mathbf{h} * \mathbf{x})}(\omega) = \hat{h}(\omega)\,\hat{x}(\omega). \tag{33}
$$

Proof: The statement follows based on the definition 2,

$$
\begin{aligned}
\hat{y}(\omega) = \widehat{(\mathbf{h} * \mathbf{x})}(\omega) &= \sum_{n=-\infty}^{\infty} y(n) e^{-\imath n\omega} = \sum_n \left(\sum_k h(k) x(n-k) \right) e^{-\imath n\omega} \\
&= \sum_k h(k) \left(\sum_n x(n-k) e^{-\imath n\omega} \right) = \sum_k h(k) \left(\sum_l x(l) e^{-\imath (l+k)\omega} \right) \\
&= \left(\sum_k h(k) e^{-\imath k\omega} \right) \left(\sum_l x(l) e^{-\imath l\omega} \right) = \hat{h}(\omega)\,\hat{x}(\omega).
\end{aligned}
$$

∎

The Fourier transform of a signal can be written in a form of the *z-transform* by substituting $z = e^{\imath\omega}$

$$
X(z) = \sum_{n=-\infty}^{\infty} x(n)\, z^{-n}. \tag{34}
$$

Thus the convolution theorem in the discrete case can be formulated in another way: *the z-transform of the convolution of two signals equals the product of z-transforms of these signals,*

$$Y(z) = H(z)\, X(z). \tag{35}$$

From the convolution theorem it follows that, for a given frequency ω, the signal $\{e^{\imath n\omega}\}_n$ is the eigenvector and the Fourier transform $\hat{h}(\omega)$ is the corresponding eigenvalue of the convolution operator defined by the signal $\mathbf{h}$. Really, if $\mathbf{x} = \{e^{\imath n\omega}\}_n$ then

$$(\mathbf{h} * \mathbf{x})(n) = \sum_k h(k) e^{\imath(n-k)\omega} = e^{\imath n\omega} \sum_k h(k) e^{-\imath k\omega} = e^{\imath n\omega}\, \hat{h}(\omega).$$

The Parseval equality for the discrete signal has the form

$$\|\mathbf{f}\|^2 = \sum_{n=-\infty}^{\infty} |f(n)|^2 = \frac{1}{2\pi} \int_{-\pi}^{\pi} |\hat{f}(\omega)|^2\, d\omega = \|\hat{f}\|^2, \tag{36}$$

being the consequence of the duality of the Fourier series and the Fourier transform discrete in time.

2.3 Fourier transform

Summing up everything that has been said in the previous section, we can distinguish the following forms of Fourier transforms

(a) Continuous Time Fourier Transform (CTFT)

$$\hat{f}(\omega) = \int_{-\infty}^{\infty} f(x)\, e^{-\imath\omega x}\, dx, \qquad \textit{Fourier transform}$$

$$f(x) = \frac{1}{2\pi} \int_{-\infty}^{\infty} \hat{f}(\omega)\, e^{\imath\omega x}\, d\omega, \qquad \textit{Inverse Fourier transform}$$

(b) Continuous Time Fourier Series (CTFS)

for a periodic function $f(x + lT) = f(x)$, $l \in \mathcal{Z}$,

$$\hat{f}(k) = \frac{1}{T} \int_{-T/2}^{T/2} f(x)\, e^{-\imath k\omega_0 x}\, dx$$

$$\omega_0 = \frac{2\pi}{T}$$

$$f(x) = \sum_{k=-\infty}^{\infty} \hat{f}(k)\, e^{\imath k\omega_0 x}$$

(c) Discrete Time Fourier Transform (DTFT)

$$\hat{f}(\omega) = \sum_{n=-\infty}^{\infty} f(n)e^{-\imath\omega n},$$

$$f(n) = \frac{1}{2\pi}\int_{-\pi}^{\pi} \hat{f}(\omega)e^{\imath\omega n}\,d\omega.$$

(d) Discrete Time Fourier Series (DTFS)
for the periodic series $f(n) = f(n + lN)$, $l \in \mathcal{Z}$

$$\hat{f}(k) = \sum_{n=0}^{N-1} f(n)(W_N)^{-nk}, \qquad k \in \mathcal{Z},$$

$$W_N = e^{\imath 2\pi/N}$$

$$f(n) = \frac{1}{N}\sum_{k=0}^{N-1} \hat{f}(k)(W_N)^{nk}, \qquad n \in \mathcal{Z},$$

Why is it necessary to perform any of the said transformations? If the signal is dependent on time, its graph will be rendered in a coordinate system of time-amplitude, where the x-axis denotes time and the y-axis the amplitude, i.e. the value of a physical variable being represented at a given moment of time. However, often the most important information is hidden in the frequency content represented by the frequency spectrum (the coefficients of the Fourier series) of the signal. It is intuitively clear that frequency correlates to the rate of change of a physical variable – if it changes fast, we say it has a high frequency, if it changes slowly, we say it has a low frequency. The Fourier transform provides the frequency content of the signal, i.e. it provides a representation in the coordinate system of frequency-amplitude. The graph of the Fourier transform shows the intensity each frequency appears with in the frequency spectrum of the signal.

To calculate the Discrete Fourier transform from formula (29) (or the inverse transform from (24)) one needs to perform N^2 multiplications of complex numbers and some additions. We say that the complexity of the algorithm is of the order $O(N^2)$. If we have in mind that length N of a signal is usually a large number, the cost of the transformation is high. The most important for practical use is to find an algorithm which will realize some transformation fast and with a low memory request. The algorithm that efficiently performs the Discrete Fourier transform is well known *Fast Fourier Transform* (FFT), proposed by Cooley and Tukey [15] in 1965. The order of complexity is $O(N \log_2 N)$, i.e. the number of multiplications almost linearly depends on the length of the signal when N is the power of 2.

FFT algorithm is based on the well known result that the discrete Fourier transform of the order N (applied to a signal of length N) can be represented by the sum of two discrete Fourier transforms of the order $N/2$. Namely, if $N = 2\,M$ we have

(37) $$W_N^2 \equiv \left(e^{\imath\frac{2\pi}{N}}\right)^2 = e^{\imath\frac{2\pi}{N/2}} = e^{\imath\frac{2\pi}{M}} \equiv W_M,$$

which gives a possibility to calculate the N-dimensional vector $\mathbf{y} = \mathcal{F}_N\,\mathbf{x}$, for $\mathbf{x} = (x_0, x_1, \ldots, x_{N-1})^\top$, from the two M-dimensional vectors

$$\mathbf{y}^e = \mathcal{F}_M \mathbf{x}^e, \qquad \mathbf{y}^o = \mathcal{F}_M \mathbf{x}^o,$$

where $\mathbf{x}^e = (x_0, x_2, \ldots, x_{N-2})^\top$ and $\mathbf{x}^o = (x_1, x_3, \ldots, x_{N-1})^\top$. The j-th component of the vector $\mathbf{y}$, considering (37), equals to

$$\begin{aligned} y_j &= \sum_{k=0}^{N-1} W_N^{kj} x_k = \sum_{k=0}^{M-1} W_N^{2kj} x_{2k} + \sum_{k=0}^{M-1} W_N^{(2k+1)j} x_{2k+1} \\ &= \sum_{k=0}^{M-1} W_M^{kj} x_k^e + W_N^j \sum_{k=0}^{M-1} W_M^{kj} x_k^o, \qquad \text{i.e.} \end{aligned} \tag{38}$$

$$y_j = y_j^e + W_N^j\, y_j^o, \qquad j = 0, \ldots, M-1.$$

The expressions for the rest of the vector $\mathbf{y}$ components

$$y_{M+j} = y_j^e - W_N^j\, y_j^o \qquad j = 0, \ldots, M-1, \tag{39}$$

we obtain when we replace j with $M+j$ in (38) and use relations

$$W_M^{k(M+j)} = W_M^{kM} W_M^{kj} = W_M^{kj}, \qquad W_N^{M+j} = W_N^{N/2}\, W_N^j = -\,W_N^j.$$

Further, this scheme is applied to calculate vectors $\mathbf{y}^e$ and $\mathbf{y}^o$ through four new $M/2$-dimensional vectors $\mathbf{x}^{ee}$, $\mathbf{x}^{eo}$, $\mathbf{x}^{oe}$ and $\mathbf{x}^{oo}$, and so on. If $N = 2^l$, where l is a natural number, we can get at the end of this process the Fourier transforms of the order 1 of the input vector $\mathbf{x}$ components,

$$y_0^{eoeee\ldots oe} = x_k, \qquad k = 0, \ldots, N-1. \tag{40}$$

The Fourier transform of the number (one-dimensional vector) is the number itself, as the Fourier matrix (27) of the order one is $\mathcal{F}_1 = (1)$.

To resume, to calculate the N-dimensional vector $\mathbf{y} = \mathcal{F}_N\,\mathbf{x}$ for a known vector $\mathbf{x}$ for $N = 2^l$ we start from the Fourier transforms (40) of the order one. We add and subtract corresponding vectors, according to formulae (38) and (39), in every step. In that way we obtain half of the number of vectors from the previous level, which are twice longer. At the end we get one vector that is equal to $\mathcal{F}_N\,\mathbf{x}$, which have to be multiplied by $1/N$ if we calculate the vector (26). To calculate vector (29) we have to replace W_N with $\overline{W}_N = W_N^{-1}$ in the above expressions. The next example presents the "butterfly" structure of the described algorithm.

EXAMPLE 3. Let us determine the FFT of vector $\mathbf{f} = \begin{pmatrix}0 & 1 & 2 & 3 & 4 & 5 & 6 & 7\end{pmatrix}^\top$. As the length of the vector is $N = 8 = 2^3$, the transformation is performed in three steps.

The calculations are presented on the Figure 2.4. The solution

$$\begin{aligned} \hat{\mathbf{f}} = (3.5 \quad &-0.5 - 1.21\imath \quad -0.5 - 0.5\imath \quad -0.5 - 0.21\imath \\ &-0.5 \quad -0.5 + 0.21\imath \quad -0.5 + 0.5\imath \quad -0.5 + 1.21\imath)^\top. \end{aligned}$$

is obtained when the numbers from the right column are divided by $N = 8$. ∎

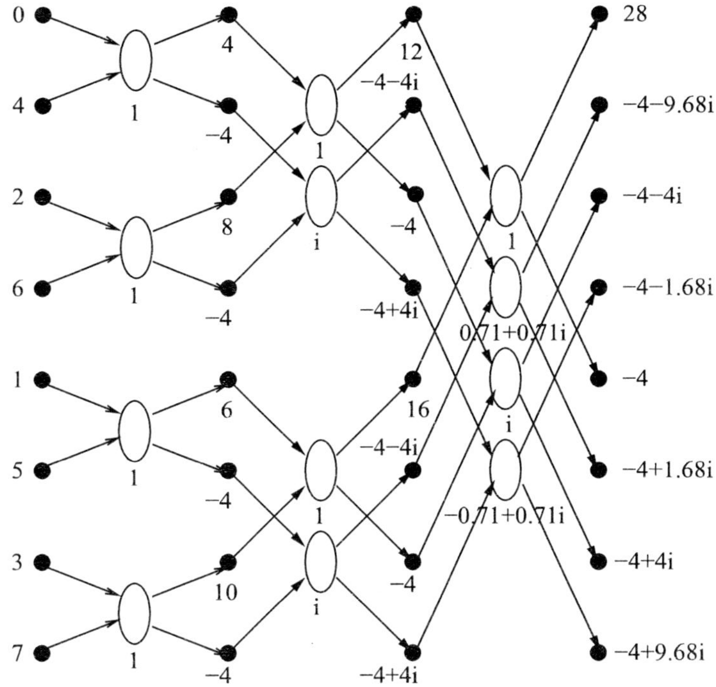

Figure 2.4: "Butterfly" structure of the FFT algorithm.

Within the Fourier transform it is not possible to localize (limit in time, if x represents time) the appearance of a harmonic in a function, because trigonometric functions have no compact supports (they are different from zero on the entire real axis). Through interference with other harmonics the effect of one frequency is canceled in a certain segment of the domain. For instance, if one tone appears in a musical theme within a limited interval of time, in the harmonic analysis of the musical signal the appropriate harmonic with the defined amplitude and phase is present, but not localized in time. Whether this tone is heard or not is regulated through interference with nearby harmonics. Therefore, the mathematical record of the musical theme through the Fourier representation is correct, but the harmonic exists in the harmonic analysis at moments when corresponding tone is not present in the musical theme as well.

Therefore we need to represent a function in the time-frequency domain, which is especially important in dealing with functions with sharp peaks or discontinuities. The Fourier analysis is not convenient for representing these functions because it provides a global representation of the function in time, and local in frequencies. A brief impulse has slowly declining Fourier spectra, thus a great number of harmonics is needed for an accurate reconstruction.

EXAMPLE 4. Dirac function $\delta(x)$ is a generalized function defined through its action to a smooth function. It represents the value of the smooth function at a point (impulse),

$$f(a) = \int_{-\infty}^{\infty} f(x)\,\delta(x-a)\,dx.$$

It can also be defined using the characteristic function $\aleph_\varepsilon(x)$ of an interval ε long, when the length of the interval tends to zero,

$$\aleph_\varepsilon(x) = \begin{cases} 1, & 0 \le x < \varepsilon \\ 0, & \text{otherwise} \end{cases}, \qquad \delta(x) = \lim_{\varepsilon \to 0} \frac{1}{\varepsilon}\,\aleph_\varepsilon(x) \quad \text{and} \quad \int_{-\infty}^{\infty} \delta(x)\,dx = 1.$$

The Fourier series (13) of Dirac function on the interval $[-\pi, \pi]$ equals

$$\begin{aligned}\delta(x) &\sim \frac{1}{2\pi}\left(1 + e^{-\imath x} + e^{\imath x} + e^{-2\imath x} + e^{2\imath x} + \cdots\right)\\ &= \frac{1}{2\pi}\left(1 + 2\cos x + 2\cos 2x + \cdots\right),\end{aligned}$$

because, according to (14),

$$c_k = \frac{1}{2\pi}\int_{-\pi}^{\pi} \delta(x) e^{-\imath k x}\,dx = \frac{1}{2\pi}, \qquad k = 0, \pm 1, \ldots .$$

Dirac function $\delta(x)$ has the Fourier coefficients $c_k = (2\pi)^{-1}$ for each k and they do not tend to zero when $k \to \infty$. The series $\sum_k |c_k|$ is divergent, whereas the Fourier series is convergent in the weak sense. The addends cancel each other out for all x except at the point $x = 0$, where the addends superimpose (Figure 2.5).

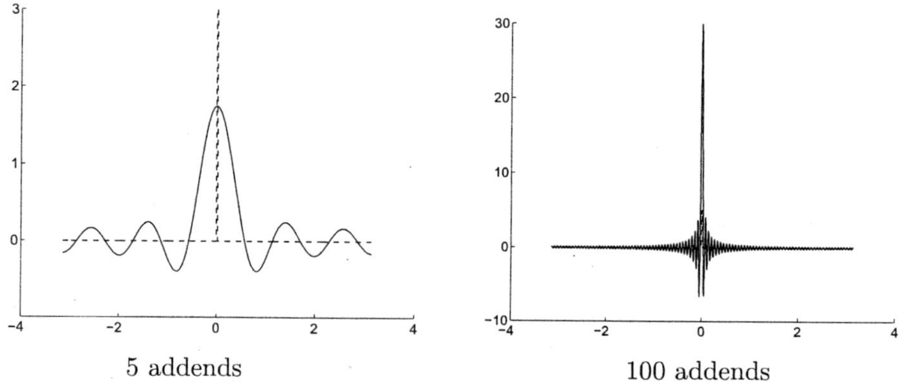

Figure 2.5: Partial sums of the Fourier series of Dirac function

It may be deduced based on the behavior of the sequence of partial sums of the Fourier series,

$$S_N(x) = \frac{1}{2\pi} \frac{\sin{(N+\frac{1}{2})x}}{\sin{\frac{1}{2}x}}, \qquad S_N(0) = \frac{2N+1}{2\pi}.$$

∎

We can see that the Fourier transform provides the spectral content of the function, but it is lacking in that it does not provide information on the moment when a component appears or disappears in time. Therefore it is useful for analyzing stationary functions, i.e. those that have spectral components that last infinitely. The Fourier transform can be used to analyze non-stationary functions only when the frequency content of the function is important, while the duration of certain harmonics is not.

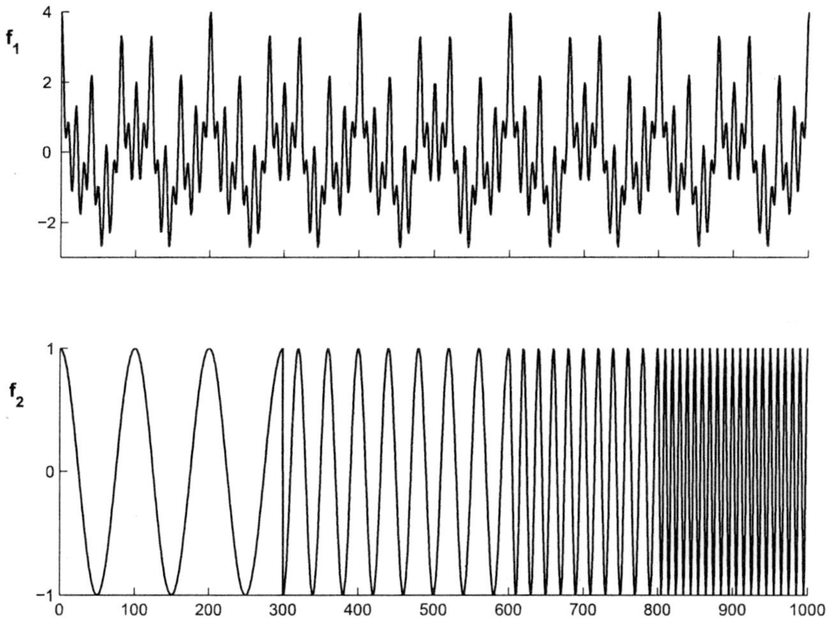

Figure 2.6: Time domain representation of a stationary (up) and a non-stationary (down) function

EXAMPLE 5. Let us compare the Fourier transforms of the stationary function (Figure 2.6, up)

$$f_1(x) = \cos{(2\pi * 10 * x)} + \cos{(2\pi * 25 * x)} + \cos{(2\pi * 50 * x)} + \cos{(2\pi * 100 * x)}$$

and the non-stationary function (Figure 2.6, down)

$$f_2(x) = \begin{cases} \cos(2\pi * 10 * x), & 0 < x < 300 \\ \cos(2\pi * 25 * x), & 300 < x < 600 \\ \cos(2\pi * 50 * x), & 600 < x < 800 \\ \cos(2\pi * 100 * x), & 800 < x < 1000 \end{cases}$$

The first function has four frequency components the whole time, whereas the second one has the same four frequency components, but at different intervals of time. The spectra of these functions are similar in form (Figure 2.7), except that smaller oscillations in the spectrum of the second function appear as a consequence of the discontinuity of the frequency. Thus, based on the Fourier transform only, it is difficult to notice the difference between these two obviously very different functions. ∎

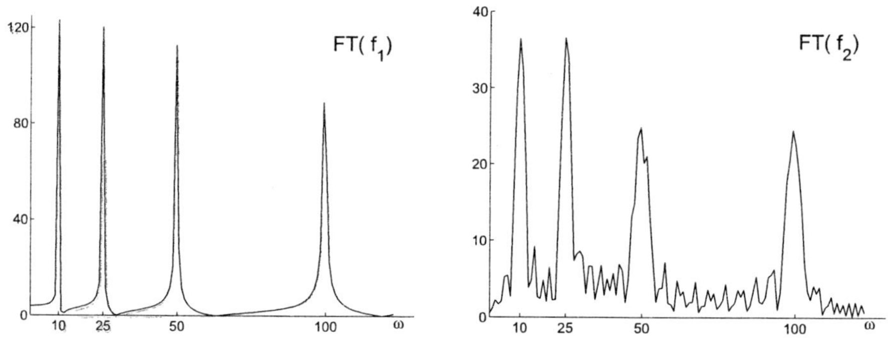

Figure 2.7: Frequency domain representations of a stationary (left) and a non-stationary (right) function

To decompose, analyze and interpret a non-stationary function and to provide information about the changes of the function spectral content through time, a time-frequency representation is needed. We can get it if we divide a function domain into small time intervals and presuming that the function is almost stationary on each of these intervals use the Fourier transform. This is the idea behind the *Short Time Fourier Transform (STFT)*. The shorter the interval the better time and the worse frequency resolution is obtained (Example 4); and vice versa, an infinite length of an interval matches a standard Fourier transform, providing a perfect frequency resolution. Segmenting the function is performed using a window function, the width of which is determined according to the length of the interval where the function is nearly stationary. The simplest window function is the characteristic function of the interval $\aleph_{(a,b)}(x)$ (see Example 2). This is not the best choice due to the discontinuity of the characteristic function. A better choice is, for example, the Gauss bell $e^{-ax^2/2}$, where a determines the width of the interval.

The STFT of a function is calculated as the Fourier transform of the product of the window function and the given function. It is obvious that this transformation is a function of the frequency, but also the time that determines the position of the window function. If we denote the window function as $w(x)$, the Short time Fourier transform of the function $f(x)$, written as $STFT_f(\omega,\tau)$ equals

$$STFT_f(\omega,\tau) = \int_{-\infty}^{\infty} f(x)\,\overline{w(x-\tau)}\,e^{-\imath\,\omega x}\,dx.$$

It is used to measure the similarity between the function on the one hand, and the shift and modulation (frequency shift) of the window function on the other,

$$STFT_f(\omega,\tau) = \big(f(x)\,,\,g_{\omega,\tau}(x)\big), \qquad g_{\omega,\tau}(x) = w(x-\tau)\,e^{\imath\,\omega x}.$$

As it is noted above, the shorter the time interval, the longer the frequency range, and vice versa. The function $\sin(x)$, representing one frequency in an infinite time interval, and the function $\delta(x)$, representing infinitely many frequencies in a single instant of time (Example 4), are extreme cases. This means that we cannot determine which frequency exists at any given moment, but merely what frequency ranges are present at various intervals of time, which is a consequence of the following statement.

UNCERTAINTY PRINCIPLE. [13] *If the function $f(x)$ decreases faster than $1/\sqrt{|x|}$ when $x \to \pm\infty$, then*

$$\Delta_x\,\Delta_\omega \geq \frac{1}{2}$$

where

(41)
$$\Delta_x = \left(\int_{-\infty}^{\infty}(x-x^*)^2\,\frac{|f(x)|^2}{\|f(x)\|^2}\,dx\right)^{1/2}, \qquad x^* = \int_{-\infty}^{\infty} x\,\frac{|f(x)|^2}{\|f(x)\|^2}\,dx,$$
$$\Delta_\omega = \left(\int_{-\infty}^{\infty}(\omega-\omega^*)^2\,\frac{|\hat f(\omega)|^2}{\|\hat f(\omega)\|^2}\,d\omega\right)^{1/2}, \qquad \omega^* = \int_{-\infty}^{\infty} \omega\,\frac{|\hat f(\omega)|^2}{\|\hat f(\omega)\|^2}\,d\omega.$$

The equality is only valid for Gauss functions $f(x) = \sqrt{\frac{a}{2\pi}}\,e^{-ax^2/2}$. ∎

The variable x^* is the center, while Δ_x is the radius of the function $f(x)$, and ω^* is the center and Δ_ω is the radius of the function $\hat f(\omega)$. If Δ_x and Δ_ω are finite values the function $f(x)$ defines a time-frequency window, or Heisenberg's box, that is represented on Figure 2.8. The surface of the window, according to said principle, is limited on the lower side. This means that an arbitrarily fine resolution (small Δ) cannot be achieved both in the temporal as well as the frequency domain. This can be stated as: it is impossible to have information on both the time and the frequency of a function in a chosen point on the time-frequency plane. The most we can find out is which spectral components exist at any given interval of time.

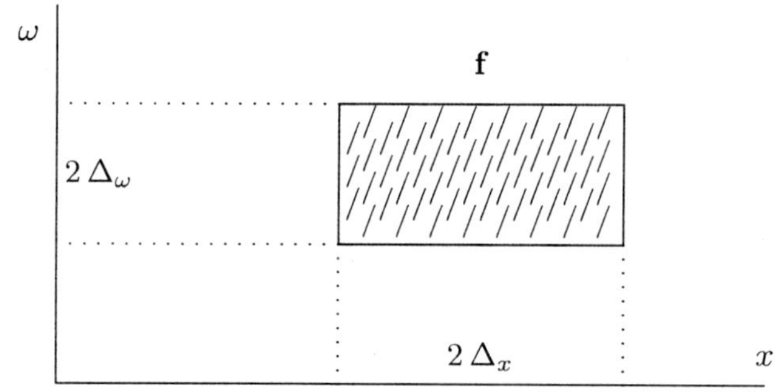

Figure 2.8: Time-frequency localization of a function

The uncertainty principle shows that it is very important how a function is divided into intervals of time in order to analyze it. The weakness of the Short time Fourier transform is that the time intervals are equal, meaning that the resolution for each x is the same. A variable time resolution enables the display of higher frequencies with a better time resolution, and the lower frequencies with a better frequency resolution. This can be achieved by defining basis functions that are interrelated by elementary transformations: *translation, modulation and scaling.* Translation is movement in time $f(x - \tau)$, whereas modulation is translation in frequency achieved by multiplying the function $f(x)$ with the function $e^{\imath\omega_0 x}$. By scaling $f(x/a)$, $a > 0$, the frequency is changed. A greater a ($a >> 1$) corresponds with long basis functions used in the analysis to describe long components of the function that change slowly. A small a ($0 < a < 1$) defines short basis functions that describe short changes.

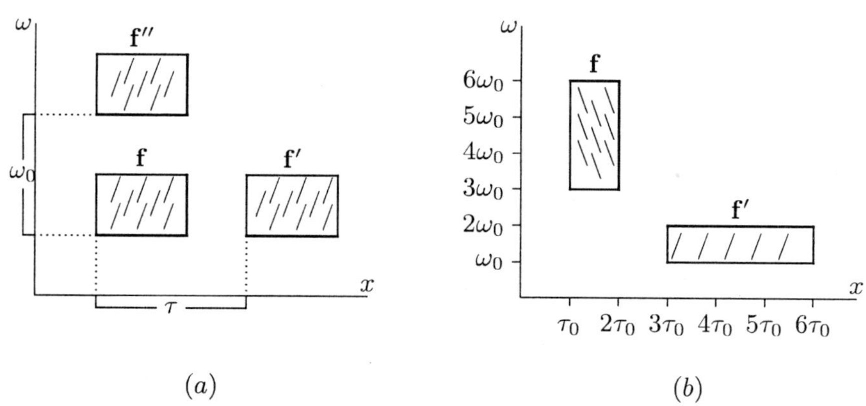

Figure 2.9: Effects of translation and modulation (a), and scaling (b)

Let us now analyze the effect of said transformations on the time-frequency rectangle **f**. The translation in time by τ produces a displacement of the rectangle by

τ along the time axis (horizontal, position $\mathbf{f}'$ on Figure 2.9(a)). Similarly, modulation by $e^{\imath\omega_0 x}$ moves the rectangle along the frequency axis (vertical, position $\mathbf{f}''$ on Figure 2.9(a)) by ω_0. Unlike these transformations, where the rectangle does not change shape, but only its position, scaling by a, $f'(x) = f(x/a)$, changes both the position and size of the rectangle $I'_x = a\, I_x$ and $I'_\omega = \frac{1}{a}\, I_\omega$ (Figure 2.9(b)) based on the scaling property of the Fourier transform.

Basis functions defined using said transformations on a single function are *wavelets.*

2.4 Wavelet transform

Dennis Gabor defined time-frequency functions, so called Gabor's wavelets, for the first time in 1946 [24]. His idea being to divide a wave, represented in mathematical notation as $\cos(\omega t + \varphi)$, into segments and then retain only one of them. Thus, Gabor's wavelet contains three pieces of information: the beginning, the end, and the frequency content in between. Difficulties arose when this transformation was to be applied to a function with a discrete argument.

A *wavelet* is a wave function with a compact support. It is called a wave due to its oscillatory nature, and diminutive is used because of the finite domain where it is different from zero (the compact support). Scaling and translation of the *basic wavelet* $\psi(x)$ (the "mother" wavelet) define the wavelet basis,

$$\psi_{a,b}(x) = \frac{1}{\sqrt{a}}\,\psi\Big(\frac{x-b}{a}\Big), \qquad a > 0.$$

By choosing appropriate values for the scaling parameter a and the translation parameter b, small segments of a complicated form may be represented with a higher resolution (zooming on sharp, brief peaks), while smooth sections can be represented with a lower resolution. It follows from the positive property of wavelets that they are basis functions limited in duration. However, the important property of Fourier basis functions $e^{\imath\omega x}$, being differential operator eigenfunctions, is lost. Wavelets are not eigenfunctions of the operator $\partial/\partial x$ and therefore frequencies are mixed meaning that in relation to the wavelet the differential operator is not diagonal. It is not possible to diagonalize the operator both in time and frequency, stemming from the above mentioned Uncertainty Principle (§2.3).

Wavelet Transform is a tool whereby data, functions or operators are being decomposed into various frequency components. Then each component is analyzed at the resolution best fit for its scale. *Continuous Wavelet Transform* (CWT) is defined by the inner product of the function and the basis wavelet,

$$CWT_f(a,b) = (f,\, \psi_{a,b}) = \frac{1}{\sqrt{a}} \int_{-\infty}^{\infty} f(x)\, \overline{\psi\Big(\frac{x-b}{a}\Big)}\, dx. \tag{42}$$

The Parseval equality stands

$$CWT_f(a,b) = \left(f\,,\,\psi_{a,b}\right) = \frac{1}{2\pi}(\hat{f}\,,\,\hat{\psi}_{a,b}),$$

where the Fourier transform of the basis wavelet is

$$\hat{\psi}_{a,b}(\omega) = \sqrt{a}\,e^{-\imath\omega b}\,\hat{\psi}(a\omega).$$

$CWT_f(a,b)$ is a function of the scale a and position b, and it shows how closely the wavelet and function correlate on the interval of time determined by the wavelet support. Wavelet transform measures the similarity between the frequency content of the function and the basis wavelet $\psi_{a,b}(x)$ in the time-frequency domain

$$\left[b + a\,(x^* - \Delta_x)\,,\, b + a\,(x^* + \Delta_x)\right] \times \left[\frac{1}{a}\,(\omega^* - \Delta_\omega)\,,\, \frac{1}{a}\,(\omega^* + \Delta_\omega)\right],$$

determined by the values (41).

Inverse Continuous Wavelet Transform

$$f(x) = \frac{1}{C_\psi}\int_{-\infty}^{\infty}\int_{-\infty}^{\infty} CWT_f(a,b)\,\psi_{a,b}(x)\,\frac{da\,db}{a^2}$$

exists if the admissibility condition is met [44]

$$C_\psi = \int_{-\infty}^{\infty}\frac{|\hat{\psi}(\omega)|^2}{\omega}\,d\omega < \infty. \tag{43}$$

Condition (43) implies that it is required that

$$\hat{\psi}(0) = \int_{-\infty}^{\infty}\psi(x)\,dx = 0, \tag{44}$$

which has as its consequence the oscillatory nature of the function $\psi(x)$. Condition (43) also implies that this function need not be equal to zero outside of the finite interval, but it must converge to zero fast enough when $|x| \to \infty$. Thus we arrive at a generalization of the wavelet concept. The basic wavelet $\psi(x)$ is an arbitrary function that has a mean value equal to zero and that decreases fast enough at infinity. These are properties of basis wavelets $\psi_{a,b}(x)$ as well. This is precisely one of the essential differences between wavelet transform and the Fourier or other transformations: while in other transformations the bases are uniquely determined, the wavelet basis is not explicitly given. The theory gives only general properties of wavelets and transformations using them, and it defines the outlines within which anyone can construct a wavelet according to their desires and requirements. When we construct or choose the basic wavelet the basis is defined with this single function. Compressed, high-frequency versions of the basic wavelet are used for temporal analysis, since details that change fast can be detected well at small time scales. Frequency analysis is performed using low-frequency dilatations of the same basic wavelet, because large scales are satisfactory for observing slow changes. These properties make wavelets an ideal tool for the analysis of non-stationary functions. Wavelet transform provides an excellent time resolution of high-frequency components and a frequency (scale) resolution of low-frequency components.

Figure 2.10 schematically compares the above mentioned transformations of a simple function, consisting of a sine wave and an impulse at the moment t_0 (Figure 2.10(a)). It is desirable to have a decomposition that includes the isolated impulse (Dirac's function per time) and the isolated frequency component (Dirac's function per frequency)

The first two decompositions (Figure 2.10 (b) and (c)) isolate the time and frequency impulse in sequence, but not both at the same time. Figure (b) represents the given function in the time domain, i.e. the decomposition is exactly $\sin x + \delta(x - t_0)$. It means that for each x the frequency is the same because the change of frequency in a single point cannot be detected. The value at point t_0 is amplified because the function equals infinity in it. Figure (c) provides a representation of the given function in the frequency domain. The Fourier spectra of the function $\delta(x)$ contains all the frequencies with the same amplitude $1/2\pi$ (Example 4), however the coefficient with the basic harmonic $\sin x$ is amplified because it is greater. The Fourier series local in time (Figure 2.10(d)) makes a compromise by locating both impulses up to a certain level. The time axis is divided into intervals of time of equal length, defining the Short Time Fourier Transform. For every interval the frequency image is given and it is the same in all intervals except in the one that contains the point t_0 where Dirac's impulse is defined. Within this interval all frequencies are present, not just the basic. The wavelet series discrete in time (Figure 2.10 (e)) provides a better localization of the time impulse, without neglecting the frequency localization. The better the localization in time, the worse it is as per frequency. Wider and shorter rectangles (on the bottom) represent low-frequency components that last longer, and narrower and taller rectangles (on the top) represent high-frequency components that last for a short time. Short, high-frequency components appear in the neighborhood of t_0. For higher frequencies the width of the rectangle becomes less, i.e. the time resolution becomes better, and the height of the rectangle increases, meaning that the frequency resolution is worse. Also, no matter the size of the rectangles, their surfaces are equal. For a single wavelet this surface is constant, while the sizes of the rectangle change with a compression or dilatation (Figure 2.9(b)). This is an effect of wavelet transform. The surface of the rectangle cannot be reduced arbitrarily by fitting a wavelet, because due to the Uncertainty Principle this surface cannot be smaller than $1/2$.

Continuous Wavelet Transform (CWT) is not of a greater practical use, because the correlation of the function and the wavelet is calculated during the wavelet being continuously translated and continuously scaled (the parameters a and b are continuous variables). Wavelets scaled in this way do not make a basis. Most of the coefficients thus calculated are redundant and there are infinitely many of them. For that reason discretization is performed, the time-scale plane is covered with a grid and at the grid nodes the CWT is calculated (for discrete values of parameters b and a). It is not the best choice to use a uniform grid because on a greater scale a

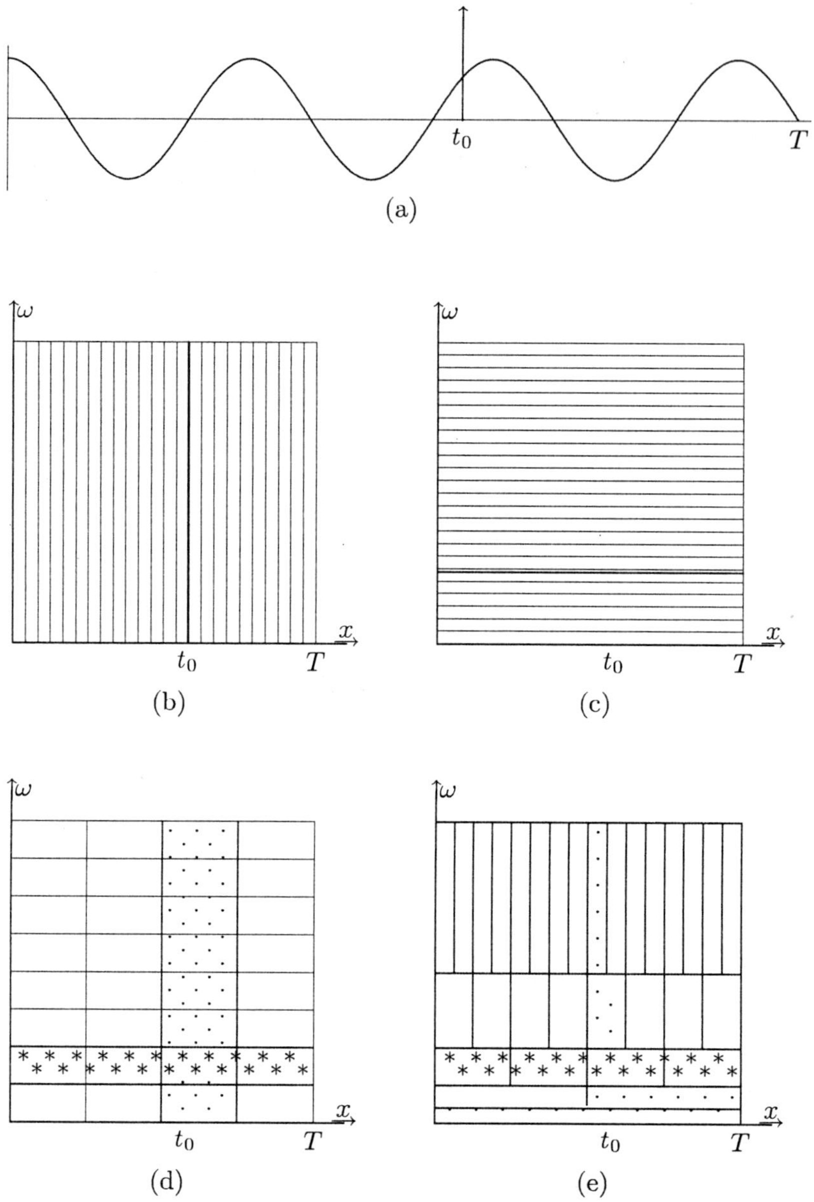

Figure 2.10: Various representations of a non-stationary function

(lower frequencies) the time step can be increased (the number of points reduced) in accordance with *Nyquist's rule* (see §6.1). This rule states that if the time-scale plane needs to contain N_1 points on the scale a_1, then it is enough to have N_2 points on the scale a_2, where

$$a_1\, N_1 = a_2\, N_2, \qquad \text{or} \qquad \frac{N_1}{\omega_1} = \frac{N_2}{\omega_2}. \tag{45}$$

Obviously, $a_1 < a_2$ means that $N_2 < N_1$. The frequency ω is equal to the inverse value of the scale, $\omega = 1/a$. For lower frequencies the number of points can be reduced, which in turn significantly reduces the number of calculations.

Fast algorithms are constructed by using discrete wavelets. *Discrete wavelets* are usually part by part continuous functions and cannot be scaled and translated continuously, but only in discrete steps,

$$\psi_{j,k}(x) = \frac{1}{\sqrt{a_0^j}}\psi\Big(\frac{x - kb_0a_0^j}{a_0^j}\Big),$$

where j and k are integers, while $a_0 > 1$ is the fixed scaling step. It is usually chosen to be $a_0 = 2$, so that the division on the frequency axis is dyadic. This is the natural choice for computers, the human ear and music, for example. The translation factor is usually chosen to be $b_0 = 1$, so that the division on the time axis of the selected scale is uniform,

$$\psi_{jk}(x) = 2^{-j/2}\psi(2^{-j}x - k), \quad \text{and} \quad \psi_{jk}(x) \neq 0 \ \text{ for } x \in [2^jk, 2^j(k+1)].$$

When the scaling parameter a is discretized on the logarithmic scale the time parameter b is discretized depending on the scaling parameter, i.e. a different number of points is used at various scales in accordance with Nyquist's rule (45) (Figure 2.11).

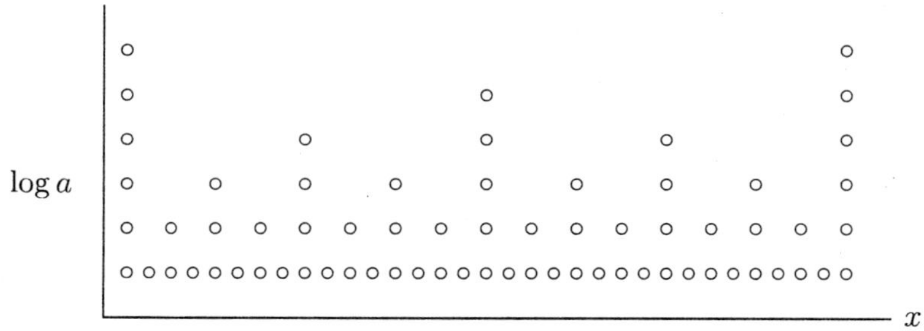

Figure 2.11: Dyadic network of points

Let us explain the procedure being described in more detail. To each point on the time-frequency plane a value of the CWT coefficient (42) is defined, thus there are infinitely many of them. We perform the discretization by defining the grid. First, the scaling parameter is discretized, by choosing integer points on the logarithmic scale - the $\log a$ axis. The base of the logarithm depends on the user, but for convenience, 2 is usually chosen. If the base of the logarithm is 2, the scales for which the transformation is to be calculated are 2, 4, 8, 16, 32, ..., and if it is, for example, 3, the scales the transformation is to be calculated for are 3, 9, 27, 81, ... After this, a uniform discretization of the time axis is performed, with its step depending on the applied discretization of the scale-axis. If the base of the logarithm chosen is two, i.e. the discrete values of the scaling parameter change for a factor of 2, the number of points on the time axis is reduced by a factor of 2 on the next scale (the step is doubled). For example, if we have 32 equidistant points on the time axis of the scale $a = 2$, then on the next scale $a = 4$ the number of points on the time axis is reduced by a factor of 2, i.e. 16 equidistant points are used and so on.

It should be noted that the discretization is arbitrary (without a limit as per the number of points on each scale) if we are only concerned with analyzing the function. If synthesizing the function is not required, even Nyquist's rule (45) need not be fulfilled. Limits in the discretization and the number of points become important if the function is expected to be reconstructed. Nyquist's number of points is the minimum number of points guaranteeing the possibility of reconstructing the continuous function (sampling theorem, §6.1). The necessary and sufficient condition for the reconstruction of a continuous function based on its series of CWT coefficients is that the energy of the wavelet coefficients is limited both on the upper and lower end, i.e. that the wavelets form a Riesz basis (10) [13]. The series of wavelets $\{\psi_{j,k}\}$ should form a basis, orthogonal or biorthogonal, or a frame (§2.1). Frames are used because it is much easier to construct them, even though this choice is not optimal. It is not easy to construct an orthogonal or biorthogonal wavelet basis.

With discrete wavelets, we still need infinitely many scalings and translations in order to calculate wavelet transform. If the function is limited in time, then the number of translations is limited as well. The matter of scaling remains – how many scalings are necessary to analyze a function? If we include the scaling function in the representation, replacing all wavelets starting from some scale J, an infinite number of wavelets in the approximation is replaced with a finite number of wavelets and scaling functions (formula (3.24)). By replacing the remaining wavelets with a finite number of scaling functions we could lose useful information, but there would be no errors in the function representation.

Using discrete wavelets, obtained by discretizing the scaling and translation parameters, is not discrete wavelet transform. The coefficients $CWT_f(a, b)$ are still determined by the integrals (42), and the function $f(x)$ is represented by a series, as a sum of the matching wavelets multiplied by the coefficients (analogy with CTFS, §2.3). To derive the discrete wavelet transformation (analogy with DTFS, §2.3) we have to discretize the algorithm for calculating wavelet coefficients. More about the discrete wavelet transform is given in §3.3.

3

Multiresolution

The wavelet transform, as defined in the previous chapter, is based on representing components of a function at different scales, i.e. with different resolutions. The idea behind the so-called multiresolution can be explained through an example taken from a real life. Geographic maps contain different information at different scales. On a global map no details can be found. Cartographers have performed a standardization of cartographic data by dividing it into independent categories matching various scales – town, region, country, continent, and globe. These categories are not entirely independent, and important data existing at a given scale are repeated on the next, larger scale. Thus, it is sufficient to determine the connections between information given on two neighboring scales (which town belongs to which region, which region belongs to which country, etc.) which can be represented through a tree-like diagram.

3.1 Multiresolution analysis

The idea of multiresolution is used in representing functions from the space $\mathcal{L}_2$.

DEFINITION 1. *Multiresolution analysis is a decomposition of the Hilbert space* $\mathcal{L}_2(R)$ *on a series of closed subspaces* $\{\mathcal{V}_j\}_{j\in Z}$ *such that*

(1) $$\cdots \subset \mathcal{V}_2 \subset \mathcal{V}_1 \subset \mathcal{V}_0 \subset \mathcal{V}_{-1} \subset \mathcal{V}_{-2} \subset \ldots$$

(2) $$\cap_{j\in Z}\mathcal{V}_j = \{0\}, \qquad \overline{\cup_{j\in Z}\mathcal{V}_j} = \mathcal{L}_2(R)$$

(3) $$\forall f \in \mathcal{L}_2(R) \ \text{ and } \ \forall j \in Z, \qquad f(x) \in \mathcal{V}_j \iff f(2x) \in \mathcal{V}_{j-1}$$

(4) $$\forall f \in \mathcal{L}_2(R) \ \text{ and } \ \forall k \in Z, \qquad f(x) \in \mathcal{V}_0 \iff f(x-k) \in \mathcal{V}_0$$

(5) $\exists \varphi \in \mathcal{V}_0$ *such that* $\{\varphi(x-k)\}_{k\in Z}$ *is a Riesz basis of the subspace* $\mathcal{V}_0$.

As a special case, in property (5) the basis can be chosen so that it is an orthonormal basis of the space $\mathcal{V}_0$.

Thus, every *approximation space* $\mathcal{V}_j$, $j \in Z$, is a scaled version of the basic space $\mathcal{V}_0$. It is obtained by dyadic scaling (by a factor of 2^j), contracting or expanding depending on the sign of j, of the space $\mathcal{V}_0$. The space $\mathcal{V}_j$, having a resolution of 2^{-j}, contains details twice as fine as those contained by its predecessor on the approximation scale $\mathcal{V}_{j+1}$. When $j \to -\infty$ the approximation becomes a representation because $\mathcal{V}_j$ tends to $\mathcal{L}_2(R)$. The basic space $\mathcal{V}_0$ is generated by one function $\varphi(x) \in \mathcal{L}_2(R)$, called the *scaling function*, because the basis of the space $\mathcal{V}_0$, according to (5), is formed by the function $\varphi(x)$ and its translations $\varphi(x-k)$, $k \in Z$. Generally, if we denote the function $\varphi(x)$, scaled j times and translated by k as

$$\varphi_{j,k}(x) = 2^{-j/2}\,\varphi(2^{-j}x - k), \qquad j,k \in Z,$$

the Riesz basis of the space $\mathcal{V}_j$ is the set of functions $\{\varphi_{j,k}(x)\}$, $k \in Z$.

Let us repeat once again (§2.3) what the effects of the elementary transformations of scaling and translation are to the function, considering that these operations are basic for multiresolution properties (3) and (4). *Dilatation* is the scaling of the function $f(x)$ by the scaling factor 2^j, written as $f(x/2^j)$. Its consequence is either the "spreading" $(j > 0)$ or "contraction" $(j < 0)$ of the function. Figure 3.1 shows the effect of dilatation to the functions $\sin x$ and the wavelet $\psi(x)$.

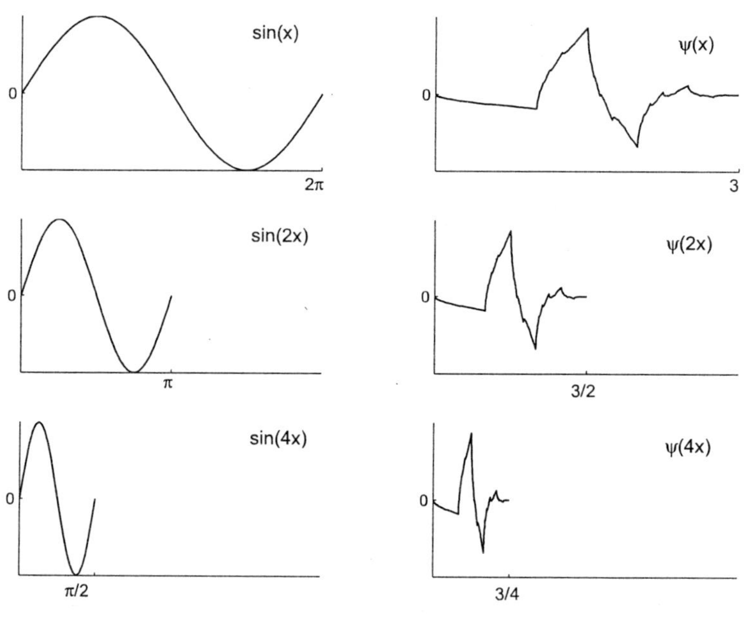

Figure 3.1: The dyadic dilatation of the sine function and the Db2 wavelet

Translation is moving the function $f(x)$ by k, written as $f(x-k)$ and its consequence is a delay ($k<0$) or speedup ($k>0$) of the function (Figure 3.2). The function $\varphi_{j,k}(x) = 2^{-j/2}\varphi(2^{-j}(x-k2^j))$ is obtained by the dyadic translation, as it is translated by $k\,2^j$.

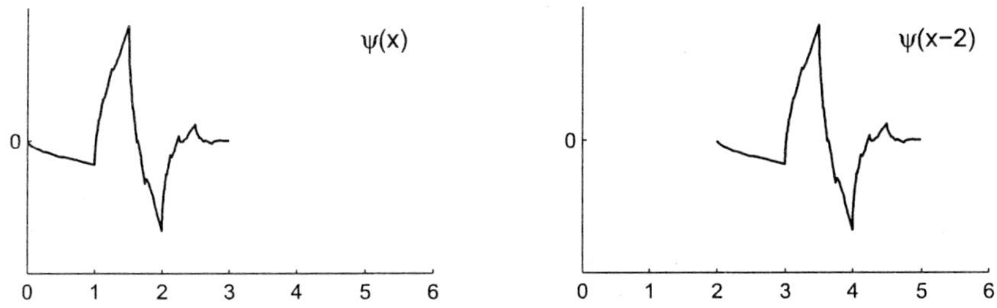

Figure 3.2: Translation of the Db2 wavelet

Moving from the space $\mathcal{V}_{j-1}$ to the space $\mathcal{V}_j$ certain details are lost due to the reduction in resolution. As $\mathcal{V}_j \subset \mathcal{V}_{j-1}$, the lost details remain conserved in the orthogonal complement of the subspace $\mathcal{V}_j$ as related to the space $\mathcal{V}_{j-1}$. This orthogonal complement is called the *wavelet space*, and we shall mark it as $\mathcal{W}_j$ on the scale j. Thus

$$\mathcal{V}_{j-1} = \mathcal{V}_j \oplus \mathcal{W}_j, \tag{6}$$

where $\oplus$ denotes the orthogonal sum. The relationship (6) yields an important property of multiresolution:

The wavelet spaces $\mathcal{W}_j$ are differences of the approximation spaces $\mathcal{V}_j$.

The approximation spaces $\mathcal{V}_j$ are sums of the wavelet spaces $\mathcal{W}_j$.

Let us explain the second statement. Based on (6), for an arbitrary J is

$$\mathcal{V}_J \oplus \mathcal{W}_J = \mathcal{V}_{J-1}, \qquad \mathcal{V}_{J-1} \oplus \mathcal{W}_{J-1} = \mathcal{V}_{J-2}.$$

By substituting the first relation in the second one, we represent the space $\mathcal{V}_{J-2}$ as a sum of three mutually orthogonal subspaces

$$\mathcal{V}_J \oplus \mathcal{W}_J \oplus \mathcal{W}_{J-1} = \mathcal{V}_{J-2}.$$

By further decomposing the approximation spaces in accordance with the same algorithm, we arrive at the space $\mathcal{V}_{j-1}$,

$$\mathcal{V}_J \oplus \mathcal{W}_J \oplus \mathcal{W}_{J-1} \oplus \cdots \oplus \mathcal{W}_j = \mathcal{V}_{j-1}, \quad J > j. \tag{7}$$

All spaces $\mathcal{W}_k$, $k \geq j$, are orthogonal to the space $\mathcal{W}_{j-1}$, because it is orthogonal to the space $\mathcal{V}_{j-1}$ which contains them. Thus, as a consequence of the relation (7) we arrive at the orthogonality of the spaces $\mathcal{W}_j$,

$$\mathcal{W}_k \perp \mathcal{W}_j, \qquad k,\, j \in Z, \quad k \neq j. \tag{8}$$

The completeness condition (2), as a limiting case of the relation (7), provides a decomposition of the space $\mathcal{L}_2(R)$. When $j \to -\infty$ we have the decomposition,

$$\mathcal{L}_2(R) = \mathcal{V}_J \oplus \sum_{j=-\infty}^{J} \mathcal{W}_j, \tag{9}$$

and another one, when $J \to \infty$,

$$\mathcal{L}_2(R) = \sum_{j=-\infty}^{\infty} \mathcal{W}_j. \tag{10}$$

Similar to the approximation spaces $\mathcal{V}_j$, the wavelet spaces $\mathcal{W}_j$ are generated by scaling and dyadic translations of another function $\psi(x) \in \mathcal{L}_2(R)$, called the *basic ("mother") wavelet*, in the sense that

$$\mathcal{W}_j = \overline{\{\psi_{j,k} \in \mathcal{L}_2 \,|\, k \in Z\}}, \qquad \psi_{j,k}(x) = 2^{-j/2}\psi(2^{-j}x - k), \quad j \in Z.$$

It needs to be emphasized that one function, *the scaling function* $\varphi(x)$ *generates the bases of each of the said spaces.* Namely, the subspaces $\mathcal{V}_0$ and $\mathcal{W}_0$ are contained in the space $\mathcal{V}_{-1}$ (relation (6) for $j = 0$), so the functions $\varphi(x) \in \mathcal{V}_0$ and $\psi(x) \in \mathcal{W}_0$ belong to the space $\mathcal{V}_{-1}$. Thus they can be represented by the basis functions $\varphi_{-1,k}(x) = \sqrt{2}\varphi(2x - k)$ of the space $\mathcal{V}_{-1}$. The matching representations are called

(11) *dilatation equation* $$\varphi(x) = \sum_k c(k)\, \sqrt{2}\, \varphi(2x - k),$$

(12) *wavelet equation* $$\psi(x) = \sum_k d(k)\, \sqrt{2}\, \varphi(2x - k).$$

With these equations the basic functions $\varphi(x)$ and $\psi(x)$ are displayed in the finer space $\mathcal{V}_{-1}$, i.e. the space with the finer scale $\Delta x = 2^{-1}$ relative to the scale $\Delta x = 1$ of the spaces $\mathcal{V}_0$ and $\mathcal{W}_0$.

The dilatation equation (11) is an equation with two scales, because it contains the function values in points x and $2x$. It has an infinite number of solutions – if $\varphi(x)$ is a solution of the equation, then the function $c\,\varphi(x)$, where c is an arbitrary constant, is also a solution. In order to define a unique solution, as far as the solution exists, we shall require the solution to be normed, i.e. that its mean value equals one,

$$\int_{-\infty}^{\infty} \varphi(x)\, dx = 1. \tag{13}$$

We mentioned that the mean value of the wavelet equals zero (formula (2.44)), determining its oscillatory nature. By integrating equation (11), taking into account condition (13),

$$1 = \int \varphi(x)\,dx = \sum_k c(k)\sqrt{2}\int \varphi(2x-k)\,dx = \frac{1}{\sqrt{2}}\sum_k c(k),$$

we get the condition to the coefficients $c(k)$ of the dilatation equation

$$\sum_k c(k) = \sqrt{2}. \tag{14}$$

The scaling function $\varphi(x)$ is entirely determined by the coefficients $c(k)$ of the dilatation equation. The development of wavelet theory was aided by noticing the relationship between coefficients $c(k)$ and filters - linear operators being used in signal processing. The desired properties of filters (and thus wavelets as well) are achieved by the appropriate choice of coefficients, to be explained in greater detail in Chapter 6.

Multiresolution can be defined by the following algorithm: starting from the coefficients $c(k)$ that meet certain properties, the dilatation equation (11) is solved by the scaling function $\varphi(x)$, which determines the basis $\{\varphi_{j,k}(x)\}$ of the approximation spaces $\mathcal{V}_j$. Then, by choosing the coefficients $d(k)$ (not always, as will be shown later, independent from the choice of coefficients $c(k)$) through expression (12) the basic wavelet $\psi(x)$ is defined, whereby the basis $\{\psi_{j,k}(x)\}$ of the wavelet space $\mathcal{W}_j$ is determined. Through dilatations and translations these spaces are automatically enlarged, completed and connected:

$$f(x) \in \mathcal{V}_j \qquad \longrightarrow \qquad f(x),\, f(2x),\, f(x-k),\, f(2x-k) \in \mathcal{V}_{j-1}.$$

Indeed, if $f(x) \in \mathcal{V}_0$, then $f(2x) \in \mathcal{V}_{-1}$, and likewise $f(2x-k) \in \mathcal{V}_{-1}$. Based on induction, it follows that $f(2^j x) \in \mathcal{V}_{-j}$ and, likewise, $f(2^j x - k) \in \mathcal{V}_{-j}$. Thus dilatation and translation are built into the $\mathcal{V}_j$ spaces. Due to translation, the domain is the entire real line $-\infty < x < \infty$, except for periodic functions. Dilatation and translation operate on the entire real axis and can be analyzed using the Fourier transform. The final condition (5) in multiresolution is related to the basis – there exists $\varphi(x)$ such that $\{\varphi(x-k)\}_k$ is an orthonormal basis of the $\mathcal{V}_0$ space. Then $\{\sqrt{2}\,\varphi(2x-k)\}_k$ is an orthonormal basis of the space $\mathcal{V}_{-1}$, and, in general, $\{2^{-j/2}\varphi(2^{-j}x-k)\}_k$ is an orthonormal basis of the space $\mathcal{V}_j$.

The one-dimensional concept of multiresolution analysis may be generalized on multidimensional spaces.

3.2 Function decomposition

The decomposition of the function space $\mathcal{L}_2(R)$ into multiresolution subspaces (formulae (9) and (10)) has, as its consequence, the possibility of decomposing an arbitrary function $f(x) \in \mathcal{L}_2(R)$, such that a component of the function $f(x)$ exists in each of those subspaces. These components, the subspace projections, contain finer and finer details of the function $f(x)$, because projections are defined on the scales $\Delta x = \ldots, 2, 1, 1/2, \ldots, 1/2^j, \ldots$.

If $\{\varphi_{j,k}(x)\}_{k\in\mathcal{Z}}$ forms an orthonormal basis of the space $\mathcal{V}_j$ we can define a projection operator in accordance with formula (2.7)

$$f_j(x) = \sum_{k\in Z} (f,\, \varphi_{j,k})\, \varphi_{j,k}(x), \qquad j \in \mathcal{Z}, \tag{15}$$

where $f_j(x)$ denotes the projection of the function $f \in \mathcal{L}_2(R)$ on the space $\mathcal{V}_j$, and (f, g) is the inner product (2.1). In relation to the previous resolution level $(j-1)$, the function approximation thus determined lost some of the fine details of the function, which can be represented by the projection operator on the complementary wavelet space $\mathcal{W}_j$,

$$\Delta f_j(x) = \sum_{k\in Z} (f,\, \psi_{j,k})\, \psi_{j,k}(x). \tag{16}$$

The coefficient $(f,\, \psi_{j,k})$ contains information on the function in the neighborhood of the point $x = 2^j k$, $\omega = 2^{-j}$ of the time-frequency plane. Lost details can be reconstructed in the approximation space on the previous level $\mathcal{V}_{j-1}$, based on relation (6),

$$f_{j-1}(x) = f_j(x) + \Delta f_j(x). \tag{17}$$

Each function $f_{j-1}(x) \in \mathcal{V}_{j-1}$ is the sum of two mutually orthogonal functions, $f_j(x)$ from $\mathcal{V}_j$ and $\Delta f_j(x)$ from $\mathcal{W}_j$. The space $\mathcal{W}_j$ contains new information $\Delta f_j(x) = f_{j-1}(x) - f_j(x)$, details at the level j. Functions from the space $\mathcal{V}_{j-1}$ can be, in accordance with relation (7), decomposed to the sum

$$\begin{aligned} f_{j-1}(x) &= f_J + (f_{J-1} - f_J) + (f_{J-2} - f_{J-1}) + \cdots + (f_{j-1} - f_j) \\ &= f_J(x) + \Delta f_J(x) + \Delta f_{J-1}(x) + \cdots + \Delta f_j(x), \end{aligned}$$

representing a multiresolution decomposition ("zooming") of the function f_{j-1}. The orthogonality of the spaces $\mathcal{W}_j$ and $\mathcal{V}_j$ is desirable, but not necessary. If spaces $\mathcal{V}_j$ and $\mathcal{W}_j$ are not orthogonal, every function $f_{j-1} \in \mathcal{V}_{j-1}$ still has a unique decomposition of the form $f_j + \Delta f_j$. This conclusion is used with so called biorthogonal wavelets, where wavelet spaces $\mathcal{W}_j$ are orthogonal to some other approximation spaces $\tilde{\mathcal{V}}_j$.

In limiting cases, based on the multiresolution completeness condition (2), it follows that $f_j(x) \to f(x)$ when $j \to -\infty$ (in the mean sense). The consequences of decompositions (9) and (10) are

$$f(x) = f_J(x) + \sum_{j=-\infty}^{J} \Delta f_j(x), \qquad f(x) = \sum_{j=-\infty}^{\infty} \Delta f_j(x). \tag{18}$$

The sums (18) represent *multiresolution decompositions* of the function $f(x)$, because they contain the basis functions at different resolution levels. Taking into consideration expressions (15) and (16), it is clear that multiresolution decomposition is characterized by two types of indices, the scaling index (resolution) j and the translation index k. The first sum contains the scaling function also, while the second contains only wavelets, but at all scales, even very large ones $\Delta x = 2^j$, $j \to \infty$. By choosing the coarsest scale J, the infinite sequence of spaces $\mathcal{W}_j, j = J+1, \ldots$, is replaced by the space $\mathcal{V}_J$, in accordance with relations (7) and (2). We can conclude that the basis of the $\mathcal{L}_2$ space is comprised of scaling functions at the chosen level J and wavelets at the levels $j \le J$, but it is also comprised of wavelets only, at all levels of resolution.

The wavelet-only decomposition often does not provide an efficient approximation of the function $f(x)$. It is far better to use a representation that contains the scaling function as well (the left sum in (18)). In order to understand this paradox, we should remember that $\varphi(x)$ is a low-frequency function with a mean value of one (condition (13)), while $\psi(x)$ is an oscillatory function with a mean value of zero (formula (2.44)). The wavelet-only approximation cannot provide a satisfactory approximation of the function that contains low-frequency spectrum components, because the scaling function is the one that adds the lost low-frequency content of the function at the coarsest resolution level.

Wavelets are usually constructed so that their moments equal zero up to a certain order

$$\int_{-\infty}^{\infty} x^l\, \psi(x)\, dx = 0, \qquad l = 0, 1, \ldots, r-1. \tag{19}$$

The number of vanishing moments r determines the width of the layer around the zero frequency where the Fourier transform of wavelets equals zero. The property of having vanishing moments up to a certain order makes wavelets very favorable for numerical modeling. In order to analyze this effect, let us look at the multiresolution decomposition of the function $f(x)$,

$$f(x) = \sum_{k\in Z} a_{J,k}\, \varphi_{J,k}(x) + \sum_{j=-\infty}^{J} \sum_{k\in Z} b_{j,k}\, \psi_{j,k}(x). \tag{20}$$

The coefficients in decomposition (20) are, based on (15), (16) and (18), Fourier coefficients according to the orthonormal function system,

$$\begin{aligned} a_{J,k} &= (f,\, \varphi_{J,k}) = \int_{\mathcal{D}_{\varphi_{j,k}}} f(x)\, \overline{\varphi_{J,k}(x)}\, dx, \\ b_{j,k} &= (f,\, \psi_{j,k}) = \int_{\mathcal{D}_{\psi_{j,k}}} f(x)\, \overline{\psi_{j,k}(x)}\, dx, \end{aligned} \tag{21}$$

where $\mathcal{D}_{\bullet_{j,k}}$ is the domain (compact support) of the matching basis function. Let us presume that $f(x)$ is a sufficiently smooth function on the wavelet domain, so that it can be represented by the Taylor polynomial

$$f(x) = \sum_{l} \alpha_l\, (x - x_m)^l \tag{22}$$

where $x_m = 2^j(m + 1/2)$ is the center of the basis wavelet. Based on (22) the second of the integrals in (21) is reduced to

$$b_{j,k} = \sum_l \alpha_l \int_{\mathcal{D}_{\psi_{j,k}}} (x - x_m)^l \, \overline{\psi_{j,k}(x)} \, dx, \tag{23}$$

and is negligible or zero if (19) is valid. The slower the function $f(x)$ oscillates on the wavelet domain, coefficients $\alpha_l \sim f^{(l)}(x_m)$ with the higher powers of Taylor's decomposition (22) are smaller, thus the wavelet coefficients $b_{j,k}$ vanish faster. As opposed to that, the wavelet coefficients will have relatively large values in areas where $f(x)$ changes rapidly or has discontinuities. Thus, a scaling function (spaces $\mathcal{V}_j$) provides information on the function $f(x)$ in an average sense while wavelets (spaces $\mathcal{W}_j$) register changes of the function $f(x)$ neglected by averaging.

In practice, the function is approximated by its rough approximation at the chosen resolution level J and details on a finite number of resolution levels j, $j \leq J$. The width of the scaling function spectrum (the number of coefficients $a_{J,k}$) is an important parameter in designing wavelet transforms. The narrower the spectrum, the greater the number of wavelet coefficients, thus more information on details is provided. Without reducing the generality, let us assume that $1 \leq j \leq J$. The projection of the function $f(x)$ onto the space $\mathcal{V}_0$ is then

$$f(x) \approx f_0(x) = \sum_{k \in Z} a_{J,k} \, \varphi_{J,k}(x) + \sum_{j=1}^{J} \sum_{k \in Z} b_{j,k} \, \psi_{j,k}(x). \tag{24}$$

The approximation error, considering (18), equals

$$\varepsilon_0(f) = f - f_0 = \sum_{j=-\infty}^{0} \Delta f_j.$$

The space $\mathcal{V}_0$ is a finite dimensional space in applications. If a function is given by a finite number of its values, the maximum number of resolution levels is determined by this number. For a function represented by 2^J of its values the maximum number of resolution levels is J.

EXAMPLE 1. [44] Let $\mathcal{V}_j$ be the space with the basis $\{e^{\imath kx}\}_{-j}^{j}$. The projection of a function $f(x)$ onto the space $\mathcal{V}_j$ is the partial sum of its Fourier series (formula (2.13)),

$$f_j(x) = \sum_{|k| \leq j} c_k \, e^{\imath kx} \qquad \text{(a part of } f(x) \text{ in } \mathcal{V}_j).$$

The basis functions are orthogonal, thus the square of the function $f_j(x)$ norm is, according to the Parseval equality (formula (2.8)), proportional to the sum $|c_k|^2$ at lower frequencies $|k| \leq j$. The error $f(x) - f_j(x)$ norm is determined by the sum of coefficients at higher frequencies $|k| > j$ and tends to zero when $j \to \infty$. Thus the sequence of spaces $\mathcal{V}_0 \cdots \subset \mathcal{V}_j \subset \mathcal{V}_{j+1} \subset \dots$ is complete in the space of 2π-periodic functions $\mathcal{L}_2(-\pi, \pi)$. But, this sequence of spaces does not generate multiresolution, because dilatation condition (3) of multiresolution is not fulfilled, $f_j(2x) \notin \mathcal{V}_{j+1}$.

If the space $\mathcal{V}_j$ is defined with the basis $\{e^{\imath kx}\}_{-2^j}^{2^j}$, the sequence of spaces $\{\mathcal{V}_j\}_0^\infty$ generates multiresolution, $f_j(2x) \in \mathcal{V}_{j+1}$. The approximation space $\mathcal{V}_j$ contains functions f_j, and the associate wavelet space $\mathcal{W}_j$ contains details $\Delta f_j(x)$,

$$f_j(x) = \sum_{|k|\le 2^j} c_k\, e^{\imath kx}, \qquad \Delta f_j(x) = \sum_{2^j<|k|\le 2^{j+1}} c_k\, e^{\imath kx}.$$

The spaces $\mathcal{V}_j$ and $\mathcal{W}_j$ have, expressed roughly, the same dimension. This is the essence of multiresolution. ∎

The decomposition (20), being performed by direct wavelet transform, is even more efficient then FFT algorithm (§2.3) if the set of basis functions $\{\varphi_{J,k},\, \psi_{j,k}\}_{j,k}$ is orthonormal. Conditions under which a scaling function and an associated wavelet generate an orthonormal basis for the space $\mathcal{L}_2$ can be expressed through coefficients of the dilatation equation (11) and the wavelet equation (12). To formulate the proposition we need to define the cascade algorithm for solving the dilatation equation. It is an iterative algorithm calculating the scaling function $\varphi(x)$, under certain conditions, as the limiting function of the function sequence

$$\varphi^{(i+1)}(x) = \sum_{k=0}^{N-1} c(k)\,\sqrt{2}\,\varphi^{(i)}(2x-k), \quad i = 0, 1, \ldots. \tag{25}$$

The cascade algorithm shall be given in more details in §4.1. The initial guess $\varphi^{(0)}(x)$ is the characteristic function of the interval $[0,\, 1]$, usually called the box function or Haar's function (see §3.4, Example 2).

THEOREM 1. *Let us assume that the cascade algorithm (25) converges uniformly, $\varphi^{(i)}(x) \to \varphi(x)$ for all x. If the coefficients $c(k)$ and $d(k)$ fulfil double shift orthogonality conditions*

$$\begin{aligned} \sum c(k)\, c(k-2m) &= \delta(m), \\ \sum d(k)\, d(k-2m) &= \delta(m), \\ \sum c(k)\, d(k-2m) &= 0, \end{aligned} \tag{26}$$

which is only possible if

$$d(k) = (-1)^k\, c(N-1-k), \qquad k = 0, \ldots, N-1, \quad N \text{ even}, \tag{27}$$

then:

(i) scaling function translations $\varphi(x-k)$ are mutually orthogonal,

$$\int_{-\infty}^{\infty} \varphi(x-n)\,\varphi(x-m)\,dx = \delta(n-m);$$

(ii) scaling functions are orthogonal to the wavelets,

$$\int_{-\infty}^{\infty} \varphi(x-n)\, \psi(x-m)\, dx = 0;$$

(iii) wavelets $\psi_{jk}(x) = 2^{-j/2}\psi(2^{-j}x - k),\ j,k \in Z,$ *form an orthonormal system,*

$$\int_{-\infty}^{\infty} \psi_{jk}(x)\, \psi_{JK}(x)\, dx = \delta(j-J)\, \delta(k-K).$$

Proof: (i) The proof shall be performed by mathematical induction. The box function is orthogonal to the translation, meaning that the functions $\varphi^{(0)}(x-n)$ comprise an orthonormal system of functions (because their supports do not overlap). Furthermore, if the functions $\varphi^{(i)}(x-n)$ comprise an orthonormal system of functions, the next iteration $\varphi^{(i+1)}(x-n)$, based on the first of the assumptions in (26), comprises an orthonormal system of functions, because

$$\begin{aligned}
&\int_{-\infty}^{\infty} \varphi^{(i+1)}(x-n)\, \varphi^{(i+1)}(x-m)\, dx \\
&= \int \Big(\sqrt{2}\sum_k c(k)\, \varphi^{(i)}(2(x-n)-k)\Big) \Big(\sqrt{2}\sum_l c(l)\, \varphi^{(i)}(2(x-m)-l)\Big)\, dx \\
&= 2\int \Big(\sum_k c(k)\, \varphi^{(i)}(2x-2n-k)\Big) \Big(\sum_{l_1} c(l_1 - 2(m-n))\, \varphi^{(i)}(2x-2n-l_1)\Big)\, dx \\
&= \sum_k \sum_{l_1} c(k)\, c(l_1 - 2(m-n)) \int \varphi^{(i)}(2(x-n)-k)\, \varphi^{(i)}(2(x-n)-l_1)\, d(2x) \\
&= \sum_k c(k)\, c(k - 2(m-n)) = \delta(m-n),
\end{aligned}$$

where the $l_1 = l - 2(n-m)$ substitution is introduced. In case of the convergence of the cascade algorithm the limiting function $\varphi(x)$ shall be orthogonal relative to the translations, too.

(ii) Similarly to (i), based on equations (11) and (12) and the third assumption in (26), it follows that

$$\begin{aligned}
&\int_{-\infty}^{\infty} \varphi(x-n)\, \psi(x-m)\, dx \\
&= \int \Big(\sqrt{2}\sum_k c(k)\, \varphi(2(x-n)-k)\Big) \Big(\sqrt{2}\sum_l d(l)\, \varphi(2(x-m)-l)\Big)\, dx \\
&= 2\int \Big(\sum_k c(k)\, \varphi(2x-2n-k)\Big) \Big(\sum_{l_1} d(l_1 - 2(m-n))\, \varphi(2x-2n-l_1)\Big)\, dx \\
&= \sum_k \sum_{l_1} c(k)\, d(l_1 - 2(m-n)) \int \varphi(2(x-n)-k)\, \varphi(2(x-n)-l_1)\, d(2x) \\
&= \sum_k c(k)\, d(k - 2(m-n)) = 0.
\end{aligned}$$

(iii) The orthogonality of wavelets at the same level of resolution (for the same j) follows from the equation (12), the orthonormality of the scaling function relative to its translations and the second assumption in (26),

$$\int_{-\infty}^{\infty} \psi(x-n)\,\psi(x-m)\,dx$$

$$= \int \Big(\sqrt{2}\sum_k d(k)\,\varphi(2(x-n)-k)\Big)\Big(\sqrt{2}\sum_l d(l)\,\varphi(2(x-m)-l)\Big)\,dx$$

$$= \sum_k d(k)\,d(k-2(m-n)) = \delta(m-n).$$

Wavelet orthogonality at various resolution levels (for different values of j) follows directly from the mutual orthogonality of the spaces $\mathcal{W}_j$, relation (8). ∎

Unlike wavelets, scaling functions are not orthogonal at various resolution levels (for different values of j). If the scaling function is orthogonal relative to its translations, the dilatation equation coefficients $c(k)$ are Fourier coefficients (formula (2.6)) of the scaling function $\varphi(x)$ by the orthonormal basis $\{\varphi_{-1,k}(x)\}$,

$$c(k) = (\varphi,\, \varphi_{-1,k}) = \int_{-\infty}^{\infty} \varphi(x)\,\sqrt{2}\varphi(2x-k)\,dx,$$

and they are not equal to zero for every k.

3.3 Pyramid algorithm

To get wavelet approximation (24) of a function $f(x) \in \mathcal{L}_2$ we apply wavelet transform. Instead of calculating the integrals (21), we start from 2^J approximation coefficients $a_{0,k}$, $k = 0, 1, \ldots, 2^J - 1$, (we shall see how to find them in §5.1) and get only by multiplications and additions the function decomposition coefficients $a_{J,k}$ and $b_{j,k}$ in J steps at most. In a single step 2^j of the approximation coefficients define 2^{j-1} new approximation coefficients and 2^{j-1} wavelet coefficients. The described procedure is called the pyramid algorithm. The reverse process is a function reconstruction and is performed by the inverse wavelet transform using an algorithm similar to the decomposition algorithm. The order of complexity of the pyramid algorithm is $O(2^J)$, i.e. the number of multiplications linearly depends on the length of the input signal.

Let us start from two subsequent multiresolution spaces $\mathcal{V}_{-1}$ and $\mathcal{V}_0$. The function $f_{-1}(x) \in \mathcal{V}_{-1}$ can be represented by the combination of the basis functions $\varphi_{-1,k}(x) = \sqrt{2}\varphi(2x-k)$ of the space $\mathcal{V}_{-1}$. But our function also can be represented by the sum of its components in spaces $\mathcal{V}_0$ and $\mathcal{W}_0$,

$$
(28)\qquad \begin{aligned}\sum_k a_{-1,k}\,\varphi_{-1,k}(x) &= \sum_k a_{0,k}\,\varphi_{0,k}(x) + \sum_k b_{0,k}\,\psi_{0,k}(x)\\ &= \sum_k a_{0,k}\,\varphi(x-k) + \sum_k b_{0,k}\,\psi(x-k),\end{aligned}
$$

as multiresolution decomposes this space into $\mathcal{V}_{-1} = \mathcal{V}_0 \oplus \mathcal{W}_0$ (formula (6)). To keep the formulae simple, we shall assume that the bases are orthonormal. In order to find the recursion, let us translate the variable x by k in the dilation equation (11) and the wavelet equation (12), and let us take $n = l - 2k$,

$$
(29)\qquad \varphi(x-k) = \sum_n c(n)\,\sqrt{2}\varphi(2x-2k-n) = \sum_l c(l-2k)\,\varphi_{-1,l}(x),
$$

$$
(30)\qquad \psi(x-k) = \sum_n d(n)\,\sqrt{2}\varphi(2x-2k-n) = \sum_l d(l-2k)\,\varphi_{-1,l}(x).
$$

Multiplication equations (29) and (30) with $f_{-1}(x)$ and integration by x gives

$$
\int f_{-1}(x)\varphi_{0,k}(x)\,dx = \int f_{-1}(x)\varphi(x-k)\,dx = \sum_l c(l-2k)\int f_{-1}(x)\varphi_{-1,l}(x)\,dx,
$$

$$
\int f_{-1}(x)\psi_{0,k}(x)\,dx = \int f_{-1}(x)\psi(x-k)\,dx = \sum_l d(l-2k)\int f_{-1}(x)\varphi_{-1,l}(x)\,dx.
$$

The bases are orthonormal, thus $a_{j,l} = (f_{-1}, \varphi_{j,l})$, $j = -1, 0$, and $b_{0,l} = (f_{-1}, \psi_{0,l})$ are the Fourier coefficients of the function $f_{-1}(x)$. By introducing these values in the above equalities, we find that the coefficients are calculated by the recursion

$$
(31)\qquad a_{0,k} = \sum_l c(l-2k)\,a_{-1,l}, \qquad b_{0,k} = \sum_l d(l-2k)\,a_{-1,l}.
$$

Generally, using recursion which makes wavelet transform fast, we determine the coefficients $a_{j,k}$ and $b_{j,k}$ based on the known coefficients $a_{j-1,k}$ during the decomposition process.

THEOREM 2. *For the function $f_{j-1}(x) = \sum_l a_{j-1,l}\,\varphi_{j-1,l}(x)$ belonging to the space $\mathcal{V}_{j-1} = \mathcal{V}_j \oplus \mathcal{W}_j$, the Fourier coefficients $a_{j,k}$ and $b_{j,k}$ by the new orthonormal basis $\{\varphi_{j,k}(x),\, \psi_{j,k}(x)\}$ are calculated using the coefficients $\mathbf{c} = \big(c(n)\big)$ and $\mathbf{d} = \big(d(n)\big)$ through the recursion*

$$
(32)\qquad a_{j,k} = \sum_l c(l-2k)\,a_{j-1,l}, \qquad b_{j,k} = \sum_l d(l-2k)\,a_{j-1,l}.
$$

Proof: : For $j = 0$ formula (32) becomes (31). The generalization for an arbitrary j follows from the dilatation equation

$$\varphi_{j,k}(x) = 2^{-j/2}\varphi(2^{-j}x - k) = 2^{-j/2}\sum_n c(n)\,\sqrt{2}\varphi(2^{-j+1}x - 2k - n)$$
$$= \sum_l c(l - 2k)\,\varphi_{j-1,l}(x).$$

The coefficients $c(n)$ need to be replaced by the coefficients $d(n)$ to get the wavelet $\psi_{j,k}(x)$ from the wavelet equation. The inner product of these functions with $f(x)$ yields the recursions (32) for the coefficients $a_{j,k}$ and $b_{j,k}$. ∎

During the process of reconstruction from the basis $\{\varphi_{j,k}(x),\, \psi_{j,k}(x)\}$ it is necessary to return to the basis $\{\varphi_{j-1,l}(x)\}$.

THEOREM 3. *The Fourier coefficients* $a_{j-1,l}$ *of a function per the orthonormal basis* $\{\varphi_{j-1,l}(x)\}$ *are calculated by the Fourier coefficients* $a_{j,k}$ *and* $b_{j,k}$ *per the orthonormal basis* $\{\varphi_{j,l}(x),\, \psi_{j,l}(x)\}$ *using the coefficients* $\mathbf{c} = \big(c(n)\big)$ *and* $\mathbf{d} = \big(d(n)\big)$ *through the recursion*

$$a_{j-1,l} = \sum_k \big(c(l - 2k)\,a_{j,k} + d(l - 2k)\,b_{j,k}\big). \tag{33}$$

Proof: For $j = 0$ the expressions (28), (29) and (30) yield the identity

$$\sum_k a_{-1,k}\,\varphi_{-1,k}(x) = \sum_n a_{0,n}\,\varphi_{0,n}(x) + \sum_n b_{0,n}\,\psi_{0,n}(x)$$
$$= \sum_n a_{0,n}\Big(\sum_l c(l - 2n)\,\varphi_{-1,l}(x)\Big) + \sum_n b_{0,n}\Big(\sum_l d(l - 2n)\,\varphi_{-1,l}(x)\Big)$$
$$= \sum_l \Big(\sum_n \big(a_{0,n}\,c(l - 2n) + b_{0,n}\,d(l - 2n)\big)\Big)\varphi_{-1,l}(x),$$

whence the statement obviously follows. For the other levels j the proof is obtained in an analogue fashion. ∎

3.4 Construction of multiresolution

Multiresolution can be constructed by defining

- the approximation spaces $\mathcal{V}_j$ and the wavelet spaces $\mathcal{W}_j$, as their differences;
- a scaling function $\varphi(x)$, translations and dilatations of which define the $\mathcal{V}_j$ spaces;

- the coefficients $c(k)$ of the dilatation equation, and by solving it we arrive at the scaling function $\varphi(x)$.

Let us demonstrate this by a few examples of multiresolution.

EXAMPLE 2. *Piecewise constant functions.* The space $\mathcal{V}_0$ is made up of all the functions from $\mathcal{L}_2$ that are constant on the unit intervals $n \le x < n+1$. Those functions are determined by their values $f(n)$ in all integer points $x = n$,

$$f(x) = f([x]), \qquad [x] \stackrel{\text{def}}{=} \text{integer part of } x.$$

The function $f(2x) \in \mathcal{V}_{-1}$ is a constant at the interval halves. The functions from $\mathcal{V}_j$ are constants on intervals 2^j in length. The spaces are increased by the reduction of j, because every function constant on the dyadic intervals $[2^j n,\, 2^j(n+1)]$ is also a constant at the halves of those intervals, so $\mathcal{V}_j \subset \mathcal{V}_{j-1}$. The spaces $\mathcal{V}_j$ are invariant relative to the translation – the translation of the piecewise constant function is still a piecewise constant function. At decrementing from j to $(j-1)$ the variable x is re-scaled by two and the $\mathcal{V}_{j-1}$ space is formed. The simplest basis is the Haar basis mentioned in the introduction. It is defined by the characteristic function of the interval $[0,1]$,

$$\varphi(x) = \begin{cases} 1, & 0 \le x < 1 \\ 0, & x \notin [0,1) \end{cases} \tag{34}$$

and we shall call this scaling function the *box function*. The box function is orthogonal relative to its translations. Every function in the space $\mathcal{V}_0$ is a linear combination of box function translations, thus can be represented as (Figure 3.3)

$$f_0(x) = \sum_n f(n)\, \varphi(x-n).$$

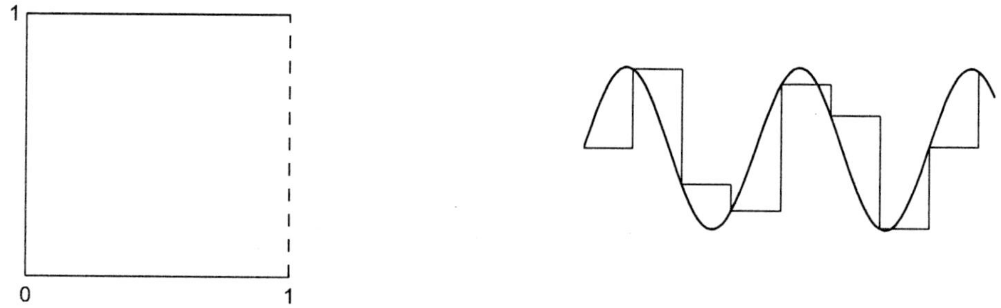

Figure 3.3: The space of piecewise constant functions

The coefficients of the dilatation equation (11) are $c(0) = c(1) = 1/\sqrt{2}$, thus the equation is

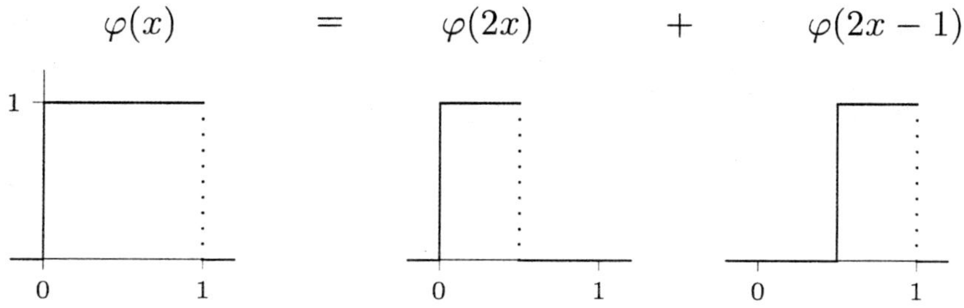

Figure 3.4: Dilatation equation of the box function

Now, let us find the corresponding wavelet. We can do this via the wavelet space $\mathcal{W}_j$. The space $\mathcal{V}_0$ contains functions constant at unit intervals, while the space $\mathcal{V}_{-1}$ contains functions constant at halves of the unit intervals. As $\mathcal{W}_0 \subset \mathcal{V}_{-1}$ the functions from $\mathcal{W}_0$ are equal to the constants at half intervals. That space is orthogonal to $\mathcal{V}_0$, because $\mathcal{V}_0 \oplus \mathcal{W}_0 = \mathcal{V}_{-1}$, which means that for arbitrary functions $g \in \mathcal{V}_0$ and $f \in \mathcal{W}_0$ the following condition should be met

$$(f, g) = \sum_k \int_k^{k+1} g(x)\, f(x)\, dx = \sum_k g(k) \int_k^{k+1} f(x)\, dx = 0.$$

In order for the above condition to be met for an arbitrary $g(k)$, $k = 0, \pm 1, \ldots$ the integral over every whole interval of the function in $\mathcal{W}_0$ must be zero,

$$\int_k^{k+1} f(x)\, dx = 0,\ k = 0, \pm 1, \ldots .$$

This fact determines the construction of the complementary space $\mathcal{W}_0$ orthogonal to $\mathcal{V}_0$ within $\mathcal{V}_{-1}$,

$$\mathcal{W}_0 = \{\textit{constant at half intervals with the condition of } f(k) + f(k + 1/2) = 0\}.$$

with the basis defined by the Haar wavelet,

$$\psi(x) = \begin{cases} 1, & x \in [0, 1/2) \\ -1, & x \in [1/2, 1) \\ 0, & x \notin [0, 1) \end{cases} \tag{35}$$

mentioned for the first time in Haar's thesis from 1909 [28]. The Haar wavelet, attached to the box function, is a function consisting from a positive and negative half box functions, defined at the first and second half of the unit interval. It is obvious that the Haar wavelet is obtained by subtracting the translated box function from the box function in the space $\mathcal{V}_{-1}$, which leads to the wavelet equation (12) (Figure 3.5).

The coefficients $d(0) = 1/\sqrt{2}$ and $d(1) = -1/\sqrt{2}$ satisfy the double shift orthogonality conditions (27) related to the box function dilatation coefficients

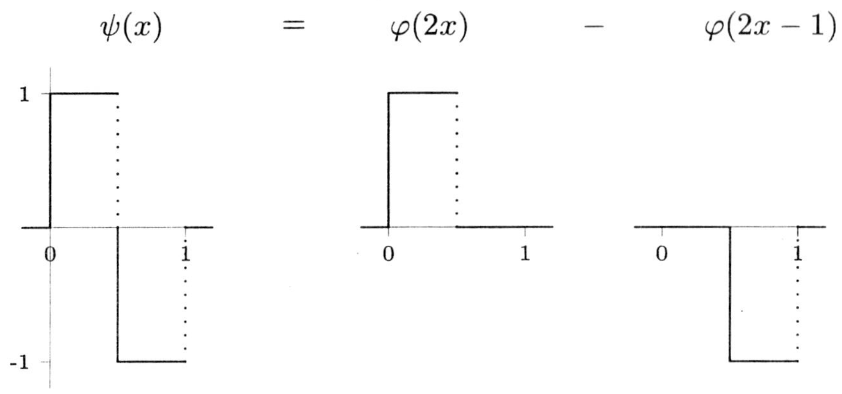

Figure 3.5: Haar wavelet equation

$c(0) = c(1) = 1/\sqrt{2}$. The Haar wavelet is orthogonal to the box function $\varphi(x)$. It is also orthogonal to the translations of the function $\varphi(x)$ and to its own translations, because the function compact supports do not overlap. Furthermore, based on multiresolution property (8), it follows that the wavelet $\psi(x)$ is orthogonal to its own dilatations and to the translations of dilatations,

$$\int_{-\infty}^{\infty} \psi(x)\, \psi(2^{-j}x - k)\, dx = 0 \qquad \text{except for} \quad j = k = 0.$$

Translations $\psi(x - k)$, $k \in \mathcal{Z}$, generate the space $\mathcal{W}_0$. Translations of the function $\psi(2^{-j}x)$ generate the space $\mathcal{W}_j$. From completeness it follows that the orthonormal system $\{\psi_{j,k}(x)\}_{j,k}$,

$$\psi_{j,k}(x) = 2^{-j/2}\, \psi(2^{-j}x - k), \qquad j, k \in \mathcal{Z},$$

is a basis in $\mathcal{L}_2$. The index j defines the resolution of the basis functions. The best least square approximation of the function $f(x)$ on this system of functions is a piecewise constant,

$$Q(x) = \sum_{j,k} (f,\, \psi_{j,k})\, \psi_{j,k}(x),$$

the values of which are equal to the mean values of the function $f(x)$ on the matching intervals. The small approximation smoothness, due to the discontinuity of the basis functions, is a drawback of Haar wavelets. ∎

EXAMPLE 3. *Continuous piecewise linear functions.* Let the basic space $\mathcal{V}_0$ be the space of functions $f(x)$ which are linear between every pair of values $f(n)$ and $f(n+1)$. These functions are invariant relative to translation because if $f(x)$ is a piecewise linear then $f(x-k)$ is a piecewise linear as well. They are also invariant relative to dilatation because if $f(x)$ is linear at unit intervals then $f(2x)$ is linear at the halves of those intervals. The basis can be defined by the translations $\varphi(x - k)$ of the linear spline, so called the *roof function*

$$\varphi(x) = \begin{cases} x, & 0 \leq x \leq 1 \\ 2 - x, & 1 \leq x \leq 2 \\ 0, & x \notin [0,2] \end{cases}$$

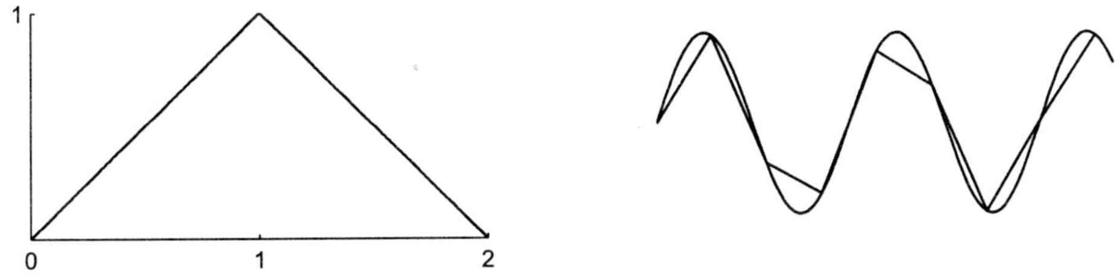

Figure 3.6: The space of continuous piecewise linear functions

Every function $f_0(x) \in \mathcal{V}_0$ can be represented by the sum $\sum f(n+1)\,\varphi(x-n)$ (Figure 3.6). The product $\varphi(x)\,\varphi(x-1)$ is positive on the interval $[1,2]$ where the compact supports of these two functions overlap, thus their inner product is not equal to zero. Therefore the roof scaling function $\varphi(x)$ is not orthogonal to it's neighboring translations $\varphi(x \pm 1)$, meaning that the basis of the roof functions is not orthogonal. Roof functions are used in the construction of biorthogonal wavelets (see §5.3).

The coefficients of the equation (11) different from zero are $c(0) = 1/(2\sqrt{2})$, $c(1) = 1/\sqrt{2}$ and $c(2) = 1/(2\sqrt{2})$, and the dilatation equation is thus (Figure 3.7)

$$\varphi(x) \quad = \quad \tfrac{1}{2}\,\varphi(2x) \quad + \quad \varphi(2x-1) \quad + \quad \tfrac{1}{2}\,\varphi(2x-2)$$

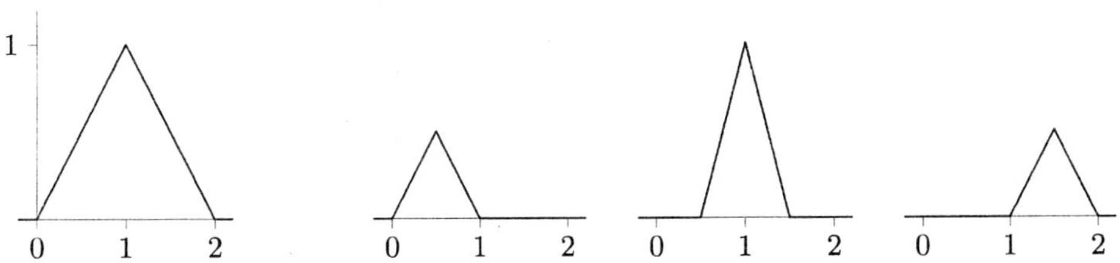

Figure 3.7: Dilatation equation of the roof function

A wavelet attached to the roof function (Figure 3.8) does not generate an orthogonal basis. The orthogonality conditions (26) cannot be fulfilled due to the odd number of $\mathbf{c}$ coefficients. ∎

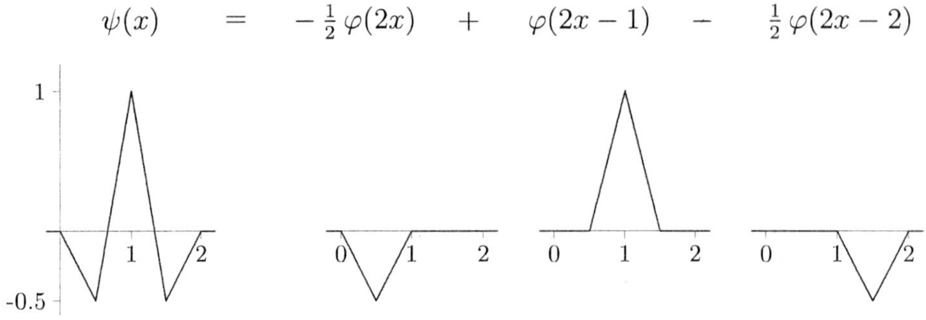

Figure 3.8: Wavelet equation of the roof function

EXAMPLE 4. *Discontinuous piecewise linear functions.* [44] Let $\mathcal{V}_0$ be the space of linear functions that can have a discontinuity at the integer points $x = n$, $n \in \mathcal{Z}$, i.e. they can have different values $f(n_-)$, when we approach the point n from the left, and $f(n_+)$, when we approach the point n from the right. The box and roof functions belong to that space (Examples 2 and 3). It is obvious that the spaces $\mathcal{V}_j$ are invariant in relation to translation and scaling. If $f(x)$ is a linear function between integers where it has a discontinuity, then it is linear between the halves of integers as well.

As there are two degrees of freedom in every integer point, the values $f(n_-)$ and $f(n_+)$, two scaling functions $\varphi_1(x)$ and $\varphi_2(x)$ are required for the construction of a basis invariant relative to the translation. Both should have a compact support on a unit interval and should be orthogonal. The natural choice is (Figure 3.9)

$$\varphi_1(x) = 1 = \text{ box function}, \qquad \varphi_2(x) = 1 - 2x = \text{line}, \qquad x \in [0, 1].$$

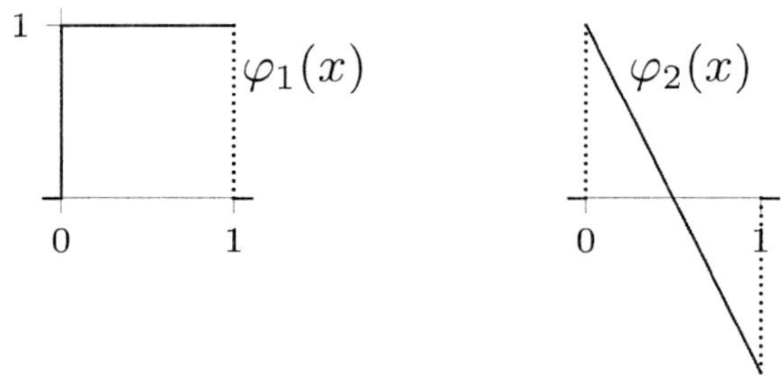

Figure 3.9: Basis functions of the discontinuous piecewise linear function space

The union of $\{\varphi_1(x-k)\}$ and $\{\varphi_2(x-k)\}$ represents an orthonormal basis illustrating the idea of *multiwavelets*. The usual dilatation equation by the scaling function $\varphi(x)$ becomes the vector equation per $\varphi_1(x)$ and $\varphi_2(x)$. The coefficients $c(k)$ in the equation are matrices 2×2 in size. ∎

EXAMPLE 5. *Cubic splines.* The space $\mathcal{V}_0$ is made up of cubic splines – functions $f(x)$ that are piecewise cubic polynomials at intervals of a unit length, where $f(x)$, $f'(x)$ and $f''(x)$ are continuous functions. The third derivative $f'''(x)$ may have a discontinuity in the integer points $x = n$, so the cubic polynomials are different at neighboring intervals. The invariance conditions shall be fulfilled in relation to translation and scaling if the space $\mathcal{V}_{-1}$ contains cubic splines at the half intervals. This condition is fulfilled based on the general rule – the approximation spaces on regular grids automatically fulfil the multiresolution conditions.

The cubic spline with the smallest support is the cubic *B-spline* (Figure 3.10). The support is made up of four unit intervals, $0 \le x \le 4$, and the representation of the cubic B-spline is as follows

$$\varphi(x) = \begin{cases} \frac{1}{6}x^3, & x \in [0,1] \\ -\frac{1}{2}(x-1)^3 + \frac{1}{2}(x-1)^2 + \frac{1}{2}(x-1) + \frac{1}{6}, & x \in [1,2] \\ \frac{1}{2}(x-2)^3 - (x-2)^2 + \frac{2}{3}, & x \in [2,3] \\ -\frac{1}{6}(x-3)^3 + \frac{1}{2}(x-3)^2 - \frac{1}{2}(x-3) + \frac{1}{6}, & x \in [3,4] \end{cases}$$

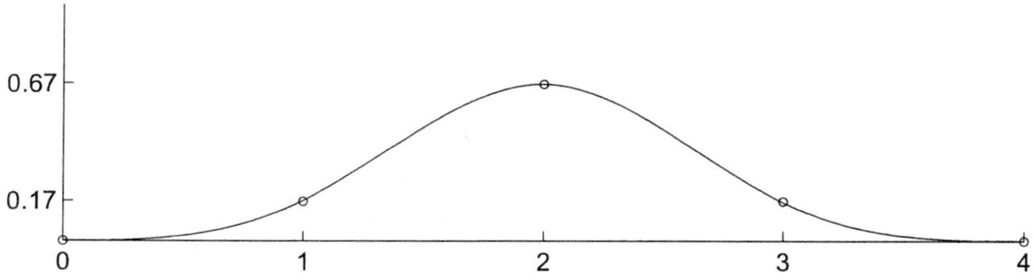

Figure 3.10: Cubic B-spline

The cubic B-spline basis is not orthogonal. Similarly to roof functions, which represent linear splines (Example 3), basis cubic splines $\varphi(x-k)$ cannot have a compact support if basis orthogonality is required. The coefficients of the dilatation equation (11) that generates the cubic B-spline are equal to scaled binomial coefficients,

$$(36) \qquad \frac{1}{8\sqrt{2}}, \quad \frac{4}{8\sqrt{2}}, \quad \frac{6}{8\sqrt{2}}, \quad \frac{4}{8\sqrt{2}}, \quad \frac{1}{8\sqrt{2}}.$$

Dividing by $8\sqrt{2}$ makes that condition (14) is fulfilled. B-splines are used in constructing biorthogonal wavelets, and more about them is given in §5.4. ▮

EXAMPLE 6. *Daubechies* Db2 *function.* In all previous examples we have started from the space $\mathcal{V}_0$. Now, we set the coefficients of the dilation equation (11)

$$(37) \qquad c(0) = \frac{1+\sqrt{3}}{4\sqrt{2}}, \quad c(1) = \frac{3+\sqrt{3}}{4\sqrt{2}}, \quad c(2) = \frac{3-\sqrt{3}}{4\sqrt{2}}, \quad c(3) = \frac{1-\sqrt{3}}{4\sqrt{2}},$$

the solution of which is the so called Daubechies scaling function $\varphi(x)$ shown on Figure 3.11, left. The functions $\varphi(x-k)$ form an orthonormal basis of the space $\mathcal{V}_0$. During the eighties of the last century Ingrid Daubechies [20] discovered an entire class of compact supported orthonormal bases linking wavelet theory with signal processing. More on this in chapter 6.

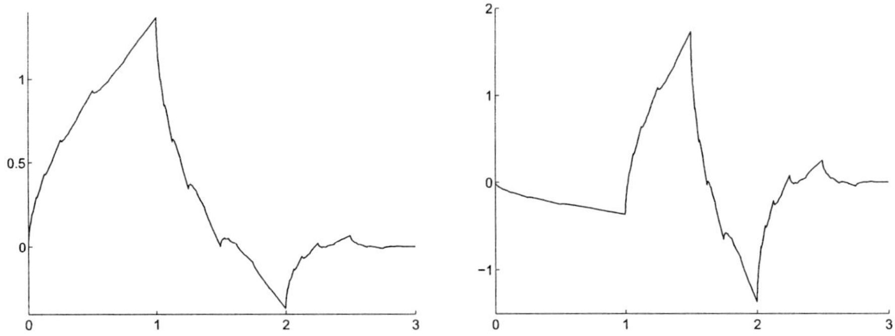

Figure 3.11: Db2 scaling function and wavelet

The Daubechies wavelet (Figure 3.11, right) is an example of constructing a wavelet from the coefficients $c(k)$. By taking the coefficients (37) in reverse order and with the alternating sign change (according to formula (27)), we arrive at four wavelet equation coefficients $d(k)$,

(38)
$$d(0) = c(3) = \frac{1-\sqrt{3}}{4\sqrt{2}}, \qquad d(1) = -\,c(2) = -\,\frac{3-\sqrt{3}}{4\sqrt{2}},$$
$$d(2) = c(1) = \frac{3+\sqrt{3}}{4\sqrt{2}}, \qquad d(3) = -\,c(0) = -\,\frac{1+\sqrt{3}}{4\sqrt{2}}.$$

Their sum is zero, while the sum of squares equals one. They are orthogonal relative to their double shift (condition (26)), because the c-s are, as well. The wavelet equation determines the Daubechies wavelet which has no explicit expression. The orthogonality of the $\psi(x-k)$ and $\varphi(x-k)$ functions is, in accordance with Theorem 1, a consequence of the orthogonality of the double shifting of the coefficients. ∎

Numbers (37) are not chosen randomly, they are coefficients of one max-flat filter used for signal processing. Daubechies constructed an entire class of orthonormal wavelet bases (see §5.2), using the analogy with the max-flat filters of signal theory, more on which will be said in §6.4. It would not have been possible to find the Daubechies wavelets Dbr by directly solving the following problem: for a given number $r \geq 0$ find the function $\psi(x) \in C^r$ such that $\{2^{-j/2}\psi(2^{-j}x-k)\}$, $j,k \in Z$, is an orthonormal basis of the space $\mathcal{L}_2(R)$.

4

Wavelets

The Fourier analysis is based on decomposing a function per sine waves with different frequencies. Similarly, the wavelet analysis is the decomposition of a function onto shifted and scaled versions of the basic wavelet. A wavelet is a wave shaped function having a limited length with a zero mean value. This means that a wavelet decreases fast enough in the frequency domain, and that

$$\hat{\psi}(0) = \int \psi(x)\, dx = 0,$$

which is a consequence of the condition for the existence of the inverse wavelet transform (condition (2.43)). Unlike a sine wave, wavelets are generally irregular and asymmetrical (Figure 4.1).

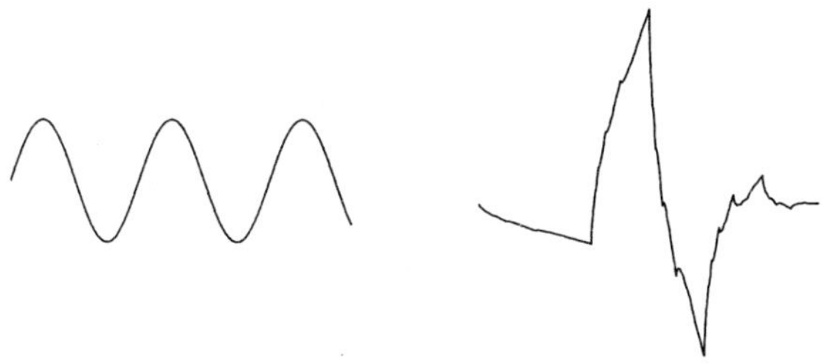

Figure 4.1: A sine wave and a wavelet

It is intuitively clear that functions with sharp changes can be analyzed better using short irregular waves than with a smooth infinite sine. The wavelet basis $\{\psi_{j,k}(x)\}_{j,k}$ is generated by the translation and dilatation $\psi(2^{-j}x - k)$ of the basic ("mother") wavelet $\psi(x)$, defined by the equation (3.12). If the basic wavelet $\psi(x)$ $(\psi(x) \equiv \psi_{0,0}(x))$ starts at the moment of $x = 0$ and ends at the moment of $x = N - 1$, the shifted wavelet $\psi_{0,k}$ starts at the moment of $x = k$ and ends at the moment of $x = k + N - 1$. The scaled wavelet $\psi_{j,0}$ starts at the moment of

$x = 0$ and ends at the moment of $x = 2^j(N-1)$. Its graph is scaled (compressed or expanded, depending of the sign of j) by a factor of 2^{-j}, while the graph of the wavelet $\psi_{0,k}$ is translated to the right by k, if $k > 0$ (Figures 3.1 and 3.2),

$$\textit{scaling} \quad \psi_{j,0}(x) = 2^{-j/2}\psi(2^{-j}x), \qquad \textit{translation} \quad \psi_{0,k}(x) = \psi(x-k).$$

The basis wavelet is generated by scaling the basic wavelet j times and shifting it by k,

$$\psi_{j,k}(x) = 2^{-j/2}\psi(2^{-j}x - k).$$

The multiplier $2^{-j/2}$ is a normalizing factor, so that the $\mathcal{L}_2$ norm of the wavelet is equal to one. The space of details on the j-th resolution level $\mathcal{W}_j$, defined in (3.6), contains functions that are linear combinations of wavelets $\psi_{jk}(x)$.

4.1 Dilatation equation

All properties of a scaling function $\varphi(x)$ and a wavelet $\psi(x)$ – such as the interval where they are different from zero, orthogonality, smoothness, vanishing moments and others, stem from properties of the coefficients of the dilatation equation (3.11) and the wavelet equation (3.12). As Theorem 3.1 proves the wavelet equation

$$(1) \qquad \psi(x) = \sqrt{2}\sum_{k=0}^{N-1}(-1)^k\, c(N-1-k)\,\varphi(2x-k), \qquad N \text{ is even},$$

defines an orthonormal basis. The alternating sign change of the wavelet equation coefficients causes the oscillatory nature of this function – thus the name wave. If the finite number of coefficients $c(k)$ is different from zero, the basic wavelet shall have a compact support – thus the diminutive.

The examples given in §3.4 show that the length of the scaling function compact support, expressed through the number of unit intervals, is determined by the number of non-zero coefficients of the dilatation equation (3.11): 2 coefficients – the length is one (the box function in Example 3.2), 3 coefficients – the length is two (the roof function in Example 3.3), 4 coefficients – the length is 3 (Daubechies function Db2 in Example 3.6), 5 coefficients – the length is four (the cubic spline in Example 3.5). If N coefficients are different from zero, $c(0), \ldots, c(N-1)$, the finite support of the function $\varphi(x)$ is the interval $[0, N-1]$, i.e. $\varphi(x)$ equals zero outside of the interval $0 \le x \le N-1$. This never happens with a single scaled difference or differential (homogenous) equation. Their solutions are expressed by the functions λ^k or $e^{\lambda x}$, where λ-s are roots of the corresponding characteristic equation. The compact support of the function $\varphi(x)$ comes from two scales in the dilatation equation. Generally, if the dilation equation has infinitely many coefficients $c(k)$, the scaling function $\varphi(x)$ has an infinite support.

THEOREM 1. *A compact support of the scaling function $\varphi(x)$ being the solution of the dilatation equation (3.11) is the interval* $[0, N-1]$.

Proof: Let us presume that we know that the support is the finite interval $[a, b]$. Then for $\varphi(2x)$ the support is the interval $[a/2,\, b/2]$. For the translated function $\varphi(2x-k)$ the support is the interval $[(a+k)/2,\, (b+k)/2]$. The index k assumes the values ranging from zero to $N-1$, so that the right side of the dilatation equation has a support within the limits of $a/2$ and $(b+N-1)/2$. Comparing it to the support of the left side of the equation we have

$$[a,\, b] = \left[\frac{a}{2},\, \frac{b+N-1}{2}\right], \qquad \text{following} \qquad a = 0, \quad b = N-1.$$

The initial assumption that the support is a finite interval follows from the cascade algorithm given below by formula (2). The initial approximation $\varphi^{(0)}(x)$ is the box function and its support is the interval $[0,\, 1]$. When it is substituted in the right side of the expression (2), we get the function $\varphi^{(1)}(x)$ which, in accordance with the previous analysis (for $a = 0$, $b = 1$) has the support $[0,\, (1+N_1)/2]$ where $N_1 = N-1$. Similarly, $\varphi^{(2)}(x)$ is zero outside of the interval $[0, (1+3N_1)/4]$ ($a = 0$, $b = (1+N_1)/2$). The function $\varphi^{(j)}(x)$ shall be zero outside of the interval $[0, \big(1+(2^j-1)N_1\big)/2^j]$, thus the boundary function, if it exists, shall be zero outside of the interval $[0,\, N_1] \equiv [0,\, N-1]$. Similarly it follows that the length of the wavelet compact support is determined by the number of non-zeroes coefficients $d(k)$ of the wavelet equation (3.12). ∎

Wavelets can be expressed using a scaling function by formula (3.12). The question to be asked first is whether the scaling function exists, i.e. whether the dilatation equation has a solution with finite energy (solution in $\mathcal{L}_2$) and how to find it? Except for trivial cases, such as Examples 3.2, 3.3 and 3.5, a scaling function as a solution of the dilatation equation (3.11) cannot be determined in analytical form. An example for this is the Daubechies scaling function given in Example 3.6. A scaling function is generally calculated by iterative algorithms whereby the function values are found on an arbitrarily dense set of dyadic points. Consequently, a wavelet associated with it is determined also on an arbitrarily dense set of dyadic points by the wavelet equation (3.12). This, however, does not reduce approximation possibilities of these functions, because there is a very efficient discrete transformation algorithm, the pyramid algorithm, which is already discussed in §3.3.

Let us now present iterative algorithms for solving dilatation equation (3.11).

Cascade algorithm was already used in §3.2 to prove orthogonality of the wavelet basis. It is an iterative algorithm for calculating a scaling function $\varphi(x)$ as the limit of function sequence $\varphi^{(j)}(x)$,

$$\varphi^{(j+1)}(x) = \sum_{k=0}^{N-1} c(k)\,\sqrt{2}\varphi^{(j)}(2x-k), \quad j = 0, 1, \ldots, \tag{2}$$

under certain constraints. The initial approximation $\varphi^{(0)}(x)$ is the box function (3.34) and its support is the interval $[0,\, 1)$. The algorithm is applied to functions with a continuous argument. Functions $\varphi^{(j)}(x)$ are piecewise constant and constant segments become shorter as the iteration index j is higher (the length of one constant segment is 2^{-j}). If the sequence $\{\varphi^{(j)}(x)\}_{j=0}^{\infty}$ converges to the limit function $\varphi(x)$, this function is the solution of the dilatation equation. Generally, convergence can be *weak* or *strong*. A weak convergence means a least square convergence, i.e. the integral of the function $\varphi^{(j)}(x)$ converges towards the integral of the function $\varphi(x)$ on every finite interval $[0,T]$ although functions $\varphi^{(j)}(x)$ can oscillate faster and faster (the integration controls the oscillations). A strong (uniform) convergence means that $\varphi^{(j)}(x)$ tends to $\varphi(x)$ at every point.

EXAMPLE 1. The box function is the fixed point of the iterative algorithm (2) defined by the coefficients $c(0) = c(1) = 1/\sqrt{2}$, because the box function is obtained in the first iteration yet,

$$\varphi^{(1)}(x) = \varphi^{(0)}(2x) + \varphi^{(0)}(2x-1) \equiv \varphi^{(0)}(x).$$

∎

EXAMPLE 2. We already know that the roof function is the solution of the dilatation equation with the coefficients $c(0) = c(2) = 1/(2\sqrt{2})$ and $c(1) = 1/\sqrt{2}$ (Example 3.3). Starting from the box function, the recursion

$$\varphi^{(1)}(x) = \frac{1}{2}\varphi^{(0)}(2x) + \varphi^{(0)}(2x-1) + \frac{1}{2}\varphi^{(0)}(2x-2)$$

defines function $\varphi^{(1)}(x)$ consisting of three box functions with half the integer interval in length (Figure 4.2, left). In the next iteration we get function $\varphi^{(2)}(x)$ consisting of seven box functions, each defined at one quarter of the integer interval (Figure 4.2, middle), etc.

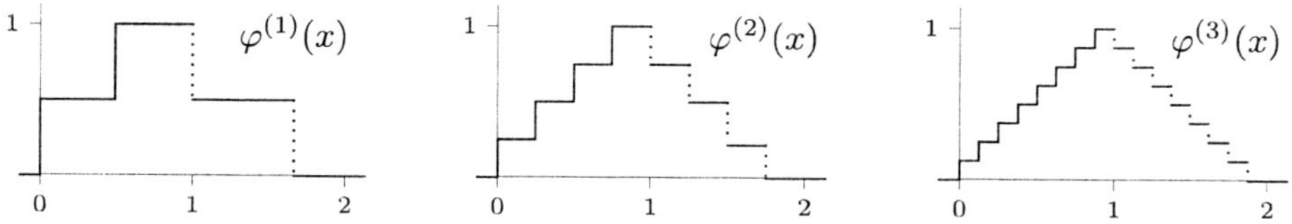

Figure 4.2: Roof function as the limit of the cascade algorithm

Scaling prevents the compact support to cross over into the infinite domain, so that the limiting interval is $[0,2]$. It is clear that the sequence of functions $\varphi^{(j)}(x)$ tends to the roof function when $j \to \infty$.

Similarly, the choice of the scaled binomial coefficients (3.36) yields to the cubic B-spline as a limiting function of the function sequence (2) (Example 3.5), while the choice of the coefficients (3.37) yields to the Daubechies scaling function (Example 3.6).

∎

The box function is chosen as an initial term of the cascade algorithm due to its orthogonality in relation to translation. A different choice of the initial function will, with strong convergence, yield a sequence of functions converging towards the same fixed point $\varphi(x)$, or the function $c\varphi(x)$, $c =$ const. A better initial approximation $\varphi^{(0)}(x)$ than the box function is a function piecewise constant on each of the intervals $n \le x < n+1$, where the value of the constant on the said interval equals the exact value of the scaling function $\varphi(n)$. These values can be determined using an algorithm based on recursion, to be explained in section §4.3.

4.2 Frequency domain

By introducing the substitution $c(k) = \sqrt{2}\, h(k)$, into the dilatation equation (3.11) it becomes

$$\varphi(x) = 2\sum_{k=0}^{N-1} h(k)\,\varphi(2x-k), \qquad \sum_{k=0}^{N-1} h(k) = 1, \tag{3}$$

whereby the condition the coefficients $h(k)$ need to satisfy is a consequence of the condition (3.14).

THEOREM 2. *The Fourier transform* $\hat\varphi(\omega)$ *of the scaling function* $\varphi(x)$ *is*

$$\hat\varphi(\omega) = \prod_{j=1}^{\infty} \hat h\bigl(\frac{\omega}{2^j}\bigr), \tag{4}$$

if the infinite product converges. The function

$$\hat h(\omega) = \sum_{k=0}^{N-1} h(k)\, e^{-\imath\omega k} \tag{5}$$

is, in signal theory, called the frequency response of the filter defined by the coefficients $h(k)$.

Proof: In order to cross over into the frequency domain, the equation (3) is multiplied by $e^{-\imath\omega x}$ and integrated by x,

$$\int_{-\infty}^{\infty} \varphi(x)\, e^{-\imath\omega x}\, dx = 2\sum_{k=0}^{N-1} h(k) \int_{-\infty}^{\infty} \varphi(2x-k)\, e^{-\imath\omega x}\, dx.$$

The left side is the Fourier transform of the scaling function $\hat\varphi(\omega)$ (formula (2.17)). By substituting $u = 2x-k$ in the integral on the right side we obtain the expression

$$2\int_{-\infty}^{\infty} \varphi(2x-k)\, e^{-\imath\omega x}\, dx = \int_{-\infty}^{\infty} \varphi(u)\, e^{-\imath\omega(u+k)/2}\, du = e^{-\imath\omega k/2}\, \hat\varphi\bigl(\frac{\omega}{2}\bigr),$$

thus the dilatation equation in the frequency domain is

$$\hat{\varphi}(\omega) = \left(\sum_{k=0}^{N-1} h(k)\, e^{-\imath\omega k/2} \right) \hat{\varphi}\big(\frac{\omega}{2}\big) = \hat{h}\big(\frac{\omega}{2}\big)\, \hat{\varphi}\big(\frac{\omega}{2}\big). \tag{6}$$

The subsequent application of relation (6) n times yields the expression

$$\hat{\varphi}(\omega) = \hat{h}\big(\frac{\omega}{2}\big)\, \hat{h}\big(\frac{\omega}{4}\big) \cdots \hat{h}\big(\frac{\omega}{2^n}\big)\, \hat{\varphi}\big(\frac{\omega}{2^n}\big) \quad \underset{n\to\infty}{\longrightarrow} \quad \hat{\varphi}(\omega) = \prod_{j=1}^{\infty} \hat{h}\big(\frac{\omega}{2^j}\big), \tag{7}$$

because $\hat{\varphi}(0) = \int \varphi(x)\, dx = 1$ (condition (3.13)). ∎

THEOREM 3. *The necessary and sufficient condition for the infinite product converges*

$$\left| \prod_{j=1}^{\infty} \hat{h}\big(\frac{\omega}{2^j}\big) \right| \le const.$$

is that the multiplier $\hat{h}(\omega/2^j) \to 1$ *when* $j \to \infty$, *i.e. that* $\hat{h}(0) = 1$.

Proof: Based on the mean value theorem and the expansion of the function $e^{|x|}$ we obtain the assessment

$$|\hat{h}(\omega)| = |1 + \hat{h}(\omega) - \hat{h}(0)| \le 1 + A|\omega| \le e^{A|\omega|}, \qquad A = \max |\hat{h}'(\omega)|,$$

therefore

$$|\hat{\varphi}(\omega)| = |\hat{h}\big(\frac{\omega}{2}\big)|\, |\hat{h}\big(\frac{\omega}{4}\big)| \cdots \le e^{A|\omega|/2}\, e^{A|\omega|/4} \cdots = e^{A|\omega| \sum_{j=1}^{\infty} 2^{-j}} = e^{A|\omega|}.$$

∎

If $\hat{h}(0) = 1$, due to the 2π-periodicity of the function $\hat{h}(\omega)$, is

$$\hat{h}(2\pi) = \hat{h}(4\pi) = \cdots = \hat{h}(2^k\pi) = \cdots = 1,$$

thus, based on (6), it follows that

$$\hat{\varphi}(2\pi) = \hat{\varphi}(4\pi) = \hat{\varphi}(8\pi) = \cdots .$$

If $\hat{h}(\pi) = 0$ as well, then all of these values are equal to zero, because

$$\hat{\varphi}(2\pi) = \hat{h}(\pi)\, \hat{\varphi}(\pi) = 0.$$

Furthermore, if $\omega = \pi$ is the zero of the order r of the function $\hat{h}(\omega)$, i.e. if $\hat{h}^{(m)}(\pi) = 0$, $m = 0, \ldots, r-1$, the Fourier transform of the scaling function $\hat{\varphi}(\omega)$ has zeroes of the order r in all points $\omega = 2n\pi$, $n = 1, 2, 3, \ldots$. Namely, for $\omega = 2\pi$ the first multiplier in product (7) is $\hat{h}(\pi)$, for $\omega = 4\pi$ the second one is $\hat{h}(\pi)$, for $\omega = 6\pi$ the first multiplier is, due to periodicity, $\hat{h}(3\pi) = \hat{h}(\pi)$, etc. Thus, the zeros of the function $\hat{\varphi}$ are the points $\omega = 2n\pi$, $n = 1, 2, \ldots$, if $\hat{h}(\pi) = 0$. The zeros of the m-th derivative of the function $\hat{\varphi}(\omega)$ are determined by the zeros up to the m-th derivative of the function $\hat{h}(\omega)$. Because $\hat{h}^{(m)}(\pi) = 0$, $m = 0, 1, \ldots, r-1$, by a recurrent procedure analogue to the above, we arrive at the conclusion that in the points $\omega = 2n\pi$, $n = 1, 2, \ldots$, the function $\hat{\varphi}(\omega)$ has zeros of the order r.

The condition $\hat{h}(\pi) = 0$ is a natural requirement in order to have the function $\hat{\varphi}(\omega)$ decreases and $\varphi(x)$ to be an "appropriate" function. We shall prove in §4.4 that the multiple zero in the point $\omega = \pi$ of the frequency response (5) is essential for good approximation properties of a wavelet basis. The property that leads to an approximation of the order r (thus the name of the condition) is

CONDITION A_r:

$$\begin{aligned} &\hat{h}(\omega) \quad \textit{has the zero of the order } r \textit{ for } \omega = \pi, \\ &\textit{i.e.} \quad \hat{h}^{(m)}(\pi) = 0, \quad m = 0, \ldots, r-1. \end{aligned}$$

It means that z-transform (2.34) of the frequency response $\hat{h}(\omega)$,

$$H(z) = \sum_{k=0}^{N-1} h(k)\, z^{-k}, \qquad z = e^{\imath\omega}, \tag{8}$$

has the zero of the order r in the point $z = -1$.

The orthogonality condition of the scaling function and its translations in the frequency domain is given by the following theorem.

THEOREM 4. *The system of functions* $\{\varphi(x-k)\}_k$ *is an orthonormal system if the Fourier transform of the scaling function* $\varphi(x)$ *meets the condition*

CONDITION O:

$$\sum_{n=-\infty}^{\infty} |\hat{\varphi}(\omega + 2n\pi)|^2 = 1.$$

Proof: Based on the generalized Parseval equality (2.23), due to the orthogonality of the system of functions $\{\varphi(x-k)\}_k$, it follows that

$$\begin{aligned} \delta(k) = \int_{-\infty}^{\infty} \varphi(x)\, \varphi(x-k)\, dx &= \frac{1}{2\pi} \int_{-\infty}^{\infty} |\hat{\varphi}(\omega)|^2\, e^{\imath k\omega}\, d\omega \\ &= \frac{1}{2\pi} \sum_{n=-\infty}^{\infty} \left(\int_{2n\pi}^{2(n+1)\pi} |\hat{\varphi}(\omega)|^2 e^{\imath k\omega}\, d\omega \right). \end{aligned}$$

By translating each of the intervals $[2n\pi, 2(n+1)\pi]$ into the basic interval $[0, 2\pi]$, the condition becomes

$$\begin{aligned} \delta(k) &= \frac{1}{2\pi} \sum_{n=-\infty}^{\infty} \left(\int_0^{2\pi} |\hat{\varphi}(\omega + 2n\pi)|^2\, e^{\imath k(\omega+2n\pi)}\, d\omega \right) \\ &= \frac{1}{2\pi} \int_0^{2\pi} \left(\sum_{n=-\infty}^{\infty} |\hat{\varphi}(\omega + 2n\pi)|^2 \right) e^{\imath k\omega}\, d\omega, \end{aligned}$$

which, for $k = 0$, is

$$\frac{1}{2\pi}\int_0^{2\pi}\left(\sum_{n=-\infty}^{\infty}|\hat{\varphi}(\omega+2n\pi)|^2\right)d\omega = 1.$$

Condition O is a direct consequence of the last expression. ∎

Let us express the wavelet in the frequency domain, now. The Fourier transform $\hat{\psi}(\omega)$ of the orthogonal wavelet fulfils the relation analogous to relation (4) for the Fourier transform of the scaling function. Indeed, the expression

$$\hat{\psi}(\omega) = \int_{-\infty}^{\infty}\psi(x)\,e^{-\imath\omega x}\,dx = \int_{-\infty}^{\infty}\left(\sqrt{2}\sum_{k=0}^{N-1}d(k)\,\varphi(2x-k)\,e^{-\imath\omega x}\right)dx$$

is, based on (3.27) for $N = 2r$, reduced to

$$\begin{aligned}\hat{\psi}(\omega) &= \sqrt{2}\sum_{k=0}^{2r-1}(-1)^k\,c(2r-1-k)\int_{-\infty}^{\infty}\varphi(2x-k)\,e^{-\imath\omega x}\,dx\\ &= \sqrt{2}\sum_{k=0}^{2r-1}(-1)^k\,c(2r-1-k)\,\frac{1}{2}\,e^{-\imath\omega k/2}\,\hat{\varphi}(\tfrac{\omega}{2}),\end{aligned}$$

i.e.

$$\hat{\psi}(\omega) = \hat{d}(\tfrac{\omega}{2})\,\hat{\varphi}(\tfrac{\omega}{2}),$$

where

(9) $$\hat{d}(\omega) = -e^{\imath(1-2r)\omega}\,\overline{\hat{h}(\omega+\pi)},$$

since

$$\begin{aligned}\hat{d}(\omega) &= \frac{1}{\sqrt{2}}\sum_{k=0}^{2r-1}(-1)^k\,c(2r-1-k)\,e^{-\imath\omega k} = \sum_{k=0}^{2r-1}(-1)^k\,h(2r-1-k)\,e^{-\imath\omega k}\\ &= \sum_{l=0}^{2r-1}(-1)^{2r-1-l}\,h(l)\,e^{-\imath\omega(2r-1-l)} = -e^{\imath(1-2r)\omega}\sum_{l=0}^{2r-1}h(l)\,e^{\imath\pi l}\,e^{\imath\omega l}\\ &= -e^{\imath(1-2r)\omega}\sum_{l=0}^{2r-1}h(l)\,e^{\imath(\omega+\pi)l}.\end{aligned}$$

Finally, considering (4), the Fourier transform of the orthogonal wavelet $\psi(x)$, with the interval $[0, 2r-1]$ as its support, equals

$$\hat{\psi}(\omega) = \hat{d}(\tfrac{\omega}{2})\,\hat{\varphi}(\tfrac{\omega}{2}) = \hat{d}(\tfrac{\omega}{2})\prod_{j=2}^{\infty}\hat{h}(\tfrac{\omega}{2^j}).$$

At the end of this section we shall illustrate by two examples how the dilatation equation can be solved in the frequency domain.

EXAMPLE 3. The coefficients $h(0) = h(1) = 1/2$ determine the box function. The frequency response (5) is $\hat{h}(\omega) = (1 + e^{-\imath\omega})/2$. The product of the first n multipliers in the expression (4) is

$$\begin{aligned}\hat{h}^{(n)}(\omega) &= \frac{1}{2^n}\,(1 + e^{-\imath\omega/2})\,(1 + e^{-\imath\omega/4})\cdots(1 + e^{-\imath\omega/2^n}) \\ &= \frac{1}{2^n}\sum_{k=0}^{2^n-1} e^{-\imath\omega k/2^n} = \frac{1 - e^{-\imath\omega}}{2^n\,(1 - e^{-\imath\omega/2^n})},\end{aligned} \tag{10}$$

where the upper limit of the sum is determined by the last addend

$$e^{-\imath\omega/2}e^{-\imath\omega/2^2}\cdots e^{-\imath\omega/2^n} = e^{-\imath\omega(\frac{1}{2}+\frac{1}{2^2}+\cdots\frac{1}{2^n})} = e^{-\imath\omega\frac{2^n-1}{2^n}} = \left(e^{-\imath\omega/2^n}\right)^{2^n-1}.$$

We analyse the behavior of the denominator in expression (10), using the Taylor's expansion for the function e^x and notation $\theta = \omega/2^n$,

$$2^n\left(1 - e^{-\imath\theta}\right) = 2^n\left(1 - (1 - \imath\theta + \cdots)\right) = 2^n\,\imath\theta + \cdots \quad \underset{n\to\infty}{\longrightarrow} \quad \imath\omega.$$

Thus, from (7) and the limit value of the product (10), we get the Fourier transform of the box function

$$\hat{\varphi}(\omega) = \prod_{j=1}^{\infty}\left(\frac{1}{2} + \frac{1}{2}e^{-\imath\omega/2^j}\right) = \lim_{n\to\infty}\hat{h}^{(n)}(\omega) = \frac{1}{\imath\omega}\,(1 - e^{-\imath\omega}) = \int_0^1 e^{-\imath\omega x}\,dx.$$

∎

EXAMPLE 4. By squaring the Fourier transform of the box function we arrive at the Fourier transform of the roof function, i.e. the linear spline, while repeated squaring yields the Fourier transform of the cubic B-spline. This is the consequence of the convolution theorem (formula 2.20) and the fact that the linear spline is the convolution of two box functions, whereas the cubic B-spline is the convolution of four box functions. More on splines in §5.4. ∎

4.3 Matrix interpretation

The dilatation equation (3) with an infinite number of coefficients can be written in vector form as,

$$\varphi(x) = 2\sum_k h(k)\,\varphi(2x - k) \qquad \Longleftrightarrow \qquad \mathbf{\Phi}_\infty(x) = M\,\mathbf{\Phi}_\infty(2x). \tag{11}$$

The vector $\mathbf{\Phi}_\infty(x) = [\ldots, \varphi(x),\ \varphi(x+1),\ \varphi(x+2), \ldots]^\top$ and the matrix

$$M = 2\begin{pmatrix} \cdot & \cdot & \cdot & \cdot & \cdot & \cdot & \\ \cdot & h(0) & 0 & 0 & 0 & 0 & \cdot \\ \cdot & h(2) & h(1) & h(0) & 0 & 0 & \cdot \\ \cdot & h(4) & h(3) & h(2) & h(1) & h(0) & \cdot \\ & \cdot & \cdot & \cdot & \cdot & \cdot & \end{pmatrix} \tag{12}$$

have infinite dimensions. All rows of the matrix M contain the coefficients $h(k)$ orderly, by having them shifted by two columns in each subsequent row, as a consequence of the argument $2x - k$. If F is the filter matrix (2.31) associated to the coefficient vector $\mathbf{h}$ and $(\downarrow 2)$ is the downsampling operator which compresses the matrix by leaving out odd rows, then

$$M = (\downarrow 2)\, 2F. \tag{13}$$

Another matrix important in wavelet theory is the matrix

$$T = (\downarrow 2)\, 2\, F\, F^\top = M\, F^\top. \tag{14}$$

The effects of the linear operators M and T to a signal f in the frequency domain are given by the expressions

$$(\hat{M f})(\omega) = \hat{h}(\frac{\omega}{2})\, \hat{f}(\frac{\omega}{2}) + \hat{h}(\frac{\omega}{2} + \pi)\, \hat{f}(\frac{\omega}{2} + \pi) \tag{15}$$

$$(\hat{T f})(\omega) = |\hat{h}(\frac{\omega}{2})|^2\, \hat{f}(\frac{\omega}{2}) + |\hat{h}(\frac{\omega}{2} + \pi)|^2\, \hat{f}(\frac{\omega}{2} + \pi). \tag{16}$$

Indeed, the Fourier transform of the vector

$$Mf = 2\begin{pmatrix} \cdot & \cdot & \cdot & \cdot & \cdot & \cdot & \cdot \\ \cdot & h(0) & 0 & 0 & 0 & 0 & \cdot \\ \cdot & h(2) & h(1) & h(0) & 0 & 0 & \cdot \\ \cdot & h(4) & h(3) & h(2) & h(1) & h(0) & \cdot \\ \cdot & \cdot & \cdot & \cdot & \cdot & \cdot & \cdot \end{pmatrix} \begin{pmatrix} \cdot \\ f(0) \\ f(1) \\ f(2) \\ \cdot \end{pmatrix}$$

$$= \begin{pmatrix} \cdot \\ 2\sum_n h(n)\, f(-n) \\ 2\sum_n h(n)\, f(2-n) \\ 2\sum_n h(n)\, f(4-n) \\ \cdot \end{pmatrix},$$

is equal to

$$(\hat{M f})(\omega) = \sum_m \left(2\sum_n h(n)\, f(2m-n)\right) e^{-\imath m\omega}.$$

On the other hand, the right side of equation (15) can be reduced to the expression

$$\hat{h}(\tfrac{\omega}{2})\,\hat{f}(\tfrac{\omega}{2}) + \hat{h}(\tfrac{\omega}{2}+\pi)\,\hat{f}(\tfrac{\omega}{2}+\pi)$$

$$= \sum_n h(n)\,e^{-\imath n\omega/2} \sum_j f(j)\,e^{-\imath j\omega/2} + \sum_n h(n)\,e^{-\imath n(\pi+\omega/2)} \sum_j f(j)\,e^{-\imath j(\pi+\omega/2)}$$

$$= \sum_{n,j} h(n)\,f(j)\,e^{-\imath(n+j)\omega/2} + \sum_{n,j} h(n)\,f(j)\,(-1)^{n+j} e^{-\imath(n+j)\omega/2}$$

$$= \sum_l (1+(-1)^l)\left(\sum_n h(n)\,f(l-n)\right) e^{-\imath l\omega/2}$$

$$= \sum_m \left(2\sum_n h(n)\,f(2m-n)\right) e^{-\imath m\omega},$$

where $l = 2m = n+j$ has been placed, because the last but one sum retains only the addends with even indices of l. This proves expression (15). Expression (16) is proven in an analogue fashion.

The iteration of the dilatation equation in the cascade algorithm (2) is equivalent to raising the matrix M to a power. Thus convergence of the cascade algorithm depends on the eigenvalues of the matrices M and T – the uniform convergence depends on the matrix M, while the least square convergence depends on the matrix T. We can get an idea about the eigenvalues of the matrix M by the following analysis. If the number of the coefficients $h(k)$ is N, the infinite-dimension problem (11) for $x=0$ is reduced to the $(N-1)$-dimension eigenvalue problem of the matrix M_{N-1},

$$\mathbf{\Phi}(0) = M_{N-1}\,\mathbf{\Phi}(0). \tag{17}$$

The matrix M_{N-1} is the square $(N-1)$-dimension block of the matrix M corresponding to the vector $\mathbf{\Phi}(0) = \big(\varphi(0), \varphi(1), \ldots, \varphi(N-2)\big)^T$. The other components of the vector $\mathbf{\Phi}_\infty(0)$ are equal to zero. The equation (17) defines the eigenvalue problem for the matrix M_{N-1}. The vector $\mathbf{\Phi}(0)$ shall be the non-zero vector if $\lambda = 1$ is the eigenvalue of the matrix M_{N-1}. Further, if $\varphi(x)$ is a differentiable function, the derivative of the dilatation equation (11)

$$\varphi'(x) = 4\sum_k h(k)\,\varphi'(2x-k) \qquad \Longleftrightarrow \qquad \mathbf{\Phi}'_\infty(x) = 2M\,\mathbf{\Phi}'_\infty(2x),$$

for $x=0$ redefines the eigenvalue problem of the matrix M_{N-1}. The vector $\mathbf{\Phi}'(0)$ shall be different from zero if $\lambda = 1/2$ is the eigenvalue of matrix M_{N-1}. By further differentiating (if possible), we arrive at a sequence of eigenvalue problems for the matrix M_{N-1},

$$\mathbf{\Phi}^{(m)}(0) = 2^m M_{N-1}\,\mathbf{\Phi}^{(m)}(0), \quad m = 0, 1, 2, \ldots .$$

The vectors $\boldsymbol{\Phi}^{(m)}(0) = \left(\varphi^{(m)}(0), \dots, \varphi^{(m)}(N-2)\right)^T$ shall be the eigenvectors of the matrix M_{N-1} if this matrix has eigenvalues $\lambda = 1/2^m$, for $m = 0, 1, 2, \dots$.

Before we proceed to analyze the eigenvalue problem for the matrix M, we shall use this matrix to arrive at the recursive algorithm for calculating the scaling function values in diadic points.

Recursion-based algorithm. If the scaling function $\varphi(x)$ is given in the integer points $x = k$ recursion (3) defines the values of this function at the halves of integers. Using the calculated values in the same way we determine the values of the function $\varphi(x)$ in quarters of integers, and generally in the diadic points $x = k/2^j$. The scaling function values at integer points are elements of an eigenvector $\boldsymbol{\Phi}(0)$. The described procedure can be efficiently realized by matrices M_0 and M_1 that are defined in the following theorem.

THEOREM 5. *The vector form of the dilation equation (11) with N addends is*

$$\boldsymbol{\Phi}(x) = M_0\, \boldsymbol{\Phi}(2x) + M_1\, \boldsymbol{\Phi}(2x-1). \tag{18}$$

The components of vector $\boldsymbol{\Phi}(x)$ are zero if the argument is outside the interval $[0,1)$. The matrices M_0 and M_1 are the $(N-1)$-dimension square blocks of matrix M with the elements

$$M_0(i,j) = M(i,j) = 2h(2i-j), \qquad M_1(i,j) = M(i,j-1) = 2h(2i-j+1)$$

The proof can be found in [44]. ∎

EXAMPLE 5. At the point $x = 0$ for $N = 5$ equation (18) is reduced to

$$\boldsymbol{\Phi}(0) = \begin{pmatrix} \varphi(0) \\ \varphi(1) \\ \varphi(2) \\ \varphi(3) \end{pmatrix} = 2 \begin{pmatrix} h(0) & 0 & 0 & 0 \\ h(2) & h(1) & h(0) & 0 \\ h(4) & h(3) & h(2) & h(1) \\ 0 & 0 & h(4) & h(3) \end{pmatrix} \begin{pmatrix} \varphi(0) \\ \varphi(1) \\ \varphi(2) \\ \varphi(3) \end{pmatrix} = M_0\, \boldsymbol{\Phi}(0).$$

When the scaling function values are determined in the integer points by solving the eigenvalue problem, values $\varphi(x)$ at integer halves are obtained by multiplying the eigenvector $\boldsymbol{\Phi}(0)$ with the matrix M_1, also stemming from (18) for $x = 1/2$,

$$\boldsymbol{\Phi}\left(\frac{1}{2}\right) = \begin{pmatrix} \varphi(1/2) \\ \varphi(3/2) \\ \varphi(5/2) \\ \varphi(7/2) \end{pmatrix} = 2 \begin{pmatrix} h(1) & h(0) & 0 & 0 \\ h(3) & h(2) & h(1) & h(0) \\ 0 & h(4) & h(3) & h(2) \\ 0 & 0 & 0 & h(4) \end{pmatrix} \begin{pmatrix} \varphi(0) \\ \varphi(1) \\ \varphi(2) \\ \varphi(3) \end{pmatrix} = M_1\, \boldsymbol{\Phi}(0).$$

Thus, the first step in the recursive algorithm for solving the dilatation equation is

$$\mathbf{\Phi}(0) = M_0\,\mathbf{\Phi}(0), \qquad \mathbf{\Phi}\big(\tfrac{1}{2}\big) = M_1\,\mathbf{\Phi}(0).$$

For every further dyadic point x the values $\varphi(x)$ are obtained through recursion by using the matrices M_0 or M_1 and values arrived at previously,

$$\mathbf{\Phi}\big(\tfrac{1}{4}\big) = M_0\,\mathbf{\Phi}\big(\tfrac{1}{2}\big), \qquad \mathbf{\Phi}\big(\tfrac{3}{4}\big) = M_1\,\mathbf{\Phi}\big(\tfrac{1}{2}\big),$$

$$\cdots \qquad \cdots \qquad \cdots$$

For the binomial coefficients

$$h(0) = \frac{1}{8}, \qquad h(1) = \frac{4}{8}, \qquad h(2) = \frac{6}{8}, \qquad h(3) = \frac{4}{8}, \qquad h(4) = \frac{1}{8},$$

we obtain the scaling function values at integer points

$$\varphi(0) = 0, \qquad \varphi(1) = \frac{1}{4}, \qquad \varphi(2) = 1, \qquad \varphi(3) = \frac{1}{4};$$

further,

$$\varphi\big(\tfrac{1}{2}\big) = \frac{1}{32}, \qquad \varphi\big(\tfrac{3}{2}\big) = \frac{23}{32}, \qquad \varphi\big(\tfrac{5}{2}\big) = \frac{23}{32}, \qquad \varphi\big(\tfrac{7}{2}\big) = \frac{1}{32}, \qquad \cdots$$

which are values of the cubic B-spline in indicated points. We already know that the cubic B-spline is the scaling function that relates to the given set of coefficients $h(k)$ (Example 3.5). ▮

The basic assumption for the existence of the non-trivial solution of the recursion-based algorithm is that matrix M_0 has the eigenvalue $\lambda = 1$, because the vector equation

$$\mathbf{\Phi}(0) = M_0\,\mathbf{\Phi}(0) \tag{19}$$

has a non-trivial solution in that case.

THEOREM 6. *If the point $\omega = \pi$ is the zero value of the filter frequency response (5), $\hat{h}(\pi) = 0$, then $\lambda = 1$ is the eigenvalue of matrices M_0, M_1 and M.*

Proof: The sum of elements in each column of the matrix M_0 equals one. Namely, when we consider that $\hat{h}(\pi) = 0$ in (5) and the filter coefficient normalizing condition (3), it follows that

$$\begin{aligned} h(0) - h(1) + h(2) - \cdots &= 0 \\ h(0) + h(1) + h(2) + \cdots &= 1 \end{aligned} \qquad \longrightarrow \qquad 2\sum_{\text{even } n} h(n) = 2\sum_{\text{odd } n} h(n) = 1,$$

i.e.

$$\sum_{\text{even } n} h(n) = \sum_{\text{odd } n} h(n) = \frac{1}{2}. \tag{20}$$

A matrix having the sum of all elements in each column equal to one has the eigenvalue $\lambda = 1$ because the vector equation (19) has a non-trivial solution. Indeed, the vector $\mathbf{e}^\top$ that has all its elements equal to one is a left eigenvector of such a matrix since

$$(1 \quad 1 \quad \cdots \quad 1)\, M_0 = (1 \quad 1 \quad \cdots \quad 1) \qquad \text{i.e.} \qquad \mathbf{e}^\top (M_0 - I) = 0.$$

The matrix $M_0 - I$ is singular and $\lambda = 1$ is the eigenvalue of the matrix M_0. This conclusion is, actually, the consequence of the basic fact that the square matrix and its transposed matrix have equal determinants, the same rank and the same eigenvalues.

The same conclusion stands for the matrices M_1 and M. ∎

It should be noted that the condition $\hat{h}(\pi) = 0$ (condition A_1 defined in §4.2) is also the necessary condition for the convergence of the cascade algorithm.

By multiplying the dilatation equation (18) from the left with the vector $\mathbf{e}^\top$, using the proven equations $\mathbf{e}^\top M_0 = \mathbf{e}^\top$ and $\mathbf{e}^\top M_1 = \mathbf{e}^\top$, we get

$$\mathbf{e}^\top \mathbf{\Phi}(x) = \mathbf{e}^\top \mathbf{\Phi}(2x) + \mathbf{e}^\top \mathbf{\Phi}(2x-1).$$

The above expression is the dilatation equation for the function $\mathbf{e}^\top \Phi(x)$, while its solution is a box function. Thus the identity

$$\sum_k \varphi(x+k) \equiv 1, \tag{21}$$

is valid.

EXAMPLE 6. The coefficients $2\,h(0) = 1/2$, $2\,h(1) = 1$ and $2\,h(2) = 1/2$ generate the roof function. The eigenvalue problem for the 2-dimension square matrix M_0 is

$$\begin{pmatrix} \frac{1}{2} & 0 \\ \frac{1}{2} & 1 \end{pmatrix} \begin{pmatrix} \varphi(0) \\ \varphi(1) \end{pmatrix} = \begin{pmatrix} \varphi(0) \\ \varphi(1) \end{pmatrix} \qquad \longrightarrow \qquad \begin{matrix} \varphi(0) = 0 \\ \varphi(1) = 1 \end{matrix}$$

The sum of all roof functions $\varphi(x+k)$ is identical to one, which is in accordance with condition (21). ∎

EXAMPLE 7. The Daubechies coefficients generate the system

$$\frac{1}{4}\begin{pmatrix} 1+\sqrt{3} & 0 & 0 \\ 3-\sqrt{3} & 3+\sqrt{3} & 1+\sqrt{3} \\ 0 & 1-\sqrt{3} & 3-\sqrt{3} \end{pmatrix} \begin{pmatrix} \varphi(0) \\ \varphi(1) \\ \varphi(2) \end{pmatrix} = \begin{pmatrix} \varphi(0) \\ \varphi(1) \\ \varphi(2) \end{pmatrix} \quad \longrightarrow \quad \begin{array}{r} \varphi(0) = 0 \\ \varphi(1) = (1+\sqrt{3})/2 \\ \varphi(2) = (1-\sqrt{3})/2 \end{array}$$

Again, the identity (21) is fulfilled. ∎

4.4 Properties

If the solution of the dilatation equation (3) exists, questions about a polynomial reconstruction, number of wavelet vanishing moments, smoothness and, finally, an approximation accuracy attained by these basis functions arise. The orthogonal basic wavelet is defined by the wavelet equation

$$\psi(x) = 2\sum_{k=0}^{N-1} (-1)^k \, h(N-1-k)\, \varphi(2x-k), \qquad N \text{ even},$$

obtained from formula (1) for $c(k) = \sqrt{2}\, h(k)$. The crucial for all answers is the multiple zero in the point $\omega = \pi$ of the frequency response $\hat{h}(\omega)$ (given in (5)),

$$\begin{aligned} &\hat{h}(\omega) \quad \textit{has a zero of the order } r \textit{ for } \omega = \pi, \\ &\textit{i.e.} \quad \hat{h}^{(m)}(\pi) = 0, \quad m = 0, \ldots, r-1, \end{aligned} \tag{22}$$

named as the CONDITION A_r, already formulated in §4.2.

The wavelet properties that are consequences of the CONDITION A_r (22) are

---

1. The dilatation matrix $M = \{2h(2i-j)\}$ has the eigenvalues $1, 1/2, \ldots, (1/2)^{r-1}$.

2. The polynomials $1, x, \ldots, x^{r-1}$ can be represented by the scaling function translations $\varphi(x-k)$, $k \in \mathcal{Z}$.

3. The first r moments of the wavelet $\psi(x)$ are equal to zero,

$$\int x^m \, \psi(x)\, dx = 0, \qquad m = 0, 1, \ldots, r-1.$$

4. A smooth function can be approximated with an error of $O((\Delta x)^r)$ through a linear combination of the scaling function translations at every level j,

$$\|f - \sum_k a_{j,k}\,\varphi_{j,k}(x)\| \le C\,(\Delta x)^r\,\|f^{(r)}\|,$$

where

$$\varphi_{j,k}(x) = 2^{-j/2}\varphi(2^{-j}x - k), \qquad a_{j,k} = \int f(x)\,\varphi_{j,k}(x)\,dx, \qquad \Delta x = 2^j.$$

5. Wavelet coefficients, i.e. Fourier coefficients of the smooth function determined by the wavelet basis decrease as

$$\int f(x)\,\psi(2^j x)\,dx \le C\,2^{-jr}.$$

6. The cascade algorithm converges in $\mathcal{L}_2$ towards the scaling function $\varphi(x)$ if the eigenvalues of the matrix T, not equal to 2^{-k}, $k = 0, \dots, 2r-1$, meet condition $|\lambda| < 1$.

7. The scaling function $\varphi(x)$ and the wavelet $\psi(x)$ have s derivatives in $\mathcal{L}_2$ if the eigenvalues of the matrix T, not equal to 2^{-k}, $k = 0, \dots, 2r-1$, meet condition $|\lambda| < 4^{-s}$ (s is not greater than the parameter r).

8. The Fourier transform of the scaling function $\hat{\varphi}(\omega)$ has zeroes of the order r in all points $\omega = 2n\pi$, $n = 1, 2, \dots$.

The property (8) has been already elaborated in §4.2. Other properties are based on the following theorems.

THEOREM 7. *Three formulations of the* CONDITION A_r *(22) are equivalent:*

(i) Frequency response (5) has the form

$$\hat{h}(\omega) = \left(\frac{1 + e^{-\imath\omega}}{2}\right)^r \hat{q}(\omega) \qquad \text{or} \qquad H(z) = \left(\frac{1 + z^{-1}}{2}\right)^r Q(z).$$

(ii) Frequency response coefficients satisfy r *summation rules*

$$\sum_{k=0}^{2r-1} (-1)^k\,k^m\,h(k) = 0, \qquad m = 0, \dots, r-1.$$

(iii) Matrix $M_{N-1} = \{2h(2i-j)\}$ *has* r *eigenvalues* $1, \frac{1}{2}, \dots, \left(\frac{1}{2}\right)^{r-1}$,

$$M_{N-1}\,\mathbf{\Phi}^{(m)}(0) = \left(\frac{1}{2}\right)^m \mathbf{\Phi}^{(m)}(0), \qquad m = 0,\, 1, \dots, r-1. \tag{23}$$

Proof: The proposition (i) follows from the factorization of a polynomial by its zeros. The proposition (ii) follows by differentiating the expression (5) m times, $m = 0, \ldots, r-1$ and calculating obtained expressions for $\omega = \pi$. The third formulation is the consequence of the following theorem. ∎

THEOREM 8. *Let us attach to the frequency response* $H(z)$ *the matrix* M_s *with the eigenvalues* λ_s *and eigenvectors* $\mathbf{x}_s$. *When we multiply* $H(z)$ *with* $\frac{1+z^{-1}}{2}$, *we get a new frequency response with the corresponding matrix* M_n *being greater by one dimension, and*

(i) *the eigenvalues* λ_n *are* $\frac{1}{2}\lambda_s$, *and the additional eigenvalue is* $\lambda_n = 1$;

(ii) *the eigenvectors* $\mathbf{x}_n$ *have coordinates which are the differences of the coordinates of the eigenvectors* $\mathbf{x}_s$,

$$x_n(k) = x_s(k) - x_s(k-1),$$

whereas in the frequency domain the link between the new and old eigenvectors is

$$X_n(z) = (1 - z^{-1})\, X_s(z).$$

Vector $\mathbf{e}^\top = \begin{pmatrix} 1 & 1 & \cdots & 1 \end{pmatrix}$ *is the left eigenvector that corresponds to the new eigenvalue* $\lambda_n = 1$. *Corresponding right eigenvector has components which are values of the scaling function at integer points* $\varphi(k)$.

The proof can be found in [44]. ∎

Now, we shall analyze the accuracy and the efficiency of the wavelet approximation (§3.2)

$$f(x) \approx f_0(x) = \sum_{k\in Z} a_{J,k}\, \varphi_{J,k}(x) + \sum_{j=1}^{J} \sum_{k\in Z} b_{j,k}\, \psi_{j,k}(x). \tag{24}$$

The accuracy of a piecewise polynomial approximation using splines or finite elements depends on the highest degree of a polynomial that can be reproduced accurately using approximation functions. When polynomials $1, x, \ldots, x^{r-1}$ belong to the approximation space, the approximation error is of the order $(\Delta x)^r$,

$$\|f(x) - f_0(x)\| \approx C(\Delta x)^r\, \|f^{(r)}(x)\|.$$

We shall prove that the approximation is mainly determined by the first addend in (24), as the scaling function translations $\varphi_{J,k}$ reconstruct polynomials. The most terms in the second addend in (24) are almost negligible because the wavelet coefficients $b_{j,k}$ are close to zero wherever the function $f(x)$ is smooth. The second addend serves to improve approximation locally, only in a part of the domain where the smoothness of the function is small.

THEOREM 9. *If the* CONDITION A_r *(22) is fulfilled, the left eigenvectors* $\mathbf{y}_m$ *of the matrix* M *given in (12), which correspond to eigenvalues* $\left(\frac{1}{2}\right)^m$,

$$\mathbf{y}_m^\top M = \Big(\frac{1}{2}\Big)^m \mathbf{y}_m^\top, \qquad m = 0, 1, \ldots, r-1,$$

determine the representation of polynomials x^m *by the scaling function and its translations* $\varphi(x+k)$,

$$x^m = \sum_k y_m(k)\, \varphi(x+k), \qquad m = 0, 1, \ldots, r-1. \tag{25}$$

Proof: Vector $\mathbf{\Phi}_\infty(x)$ with components $\varphi(x+k)$ is the solution of the dilatation equation (11). Therefore the inner product

$$\eta(x) = \mathbf{y}_m^\top \mathbf{\Phi}_\infty(x) = \sum y_m(k)\, \varphi(x+k),$$

satisfies the equation

$$\eta(x) = \mathbf{y}_m^\top \mathbf{\Phi}_\infty(x) = \mathbf{y}_m^\top M\, \mathbf{\Phi}_\infty(2x) = \left(\frac{1}{2}\right)^m \mathbf{y}_m^\top \mathbf{\Phi}_\infty(2x) = \left(\frac{1}{2}\right)^m \eta(2x).$$

Its solution is equal to a multiple of x^m because equality $\eta(2x) = 2^m\, \eta(x)$ is valid for every x only if $\eta(x) = c\,x^m$, c is an arbitrary constant.

The polynomials $1, \ldots, x^{r-1}$ do not belong to the space $\mathcal{V}_0$ because, as their support is not finite, they have infinite energy $\int_{-\infty}^{\infty} (x^j)^2\, dx = \infty$. Still, they can be represented accurately by the scaling function and its translations at every finite interval. The eigenvectors $\mathbf{y}_m$ have infinitely many components different from zero, multiplying all the translations of the function $\varphi(x)$, so that the combination remains a polynomial for every x. Thus the space $\mathcal{V}_0$ generated by the functions $\{\varphi(x-k)\}$, in a way, contains all the polynomials with a degree less than r.

∎

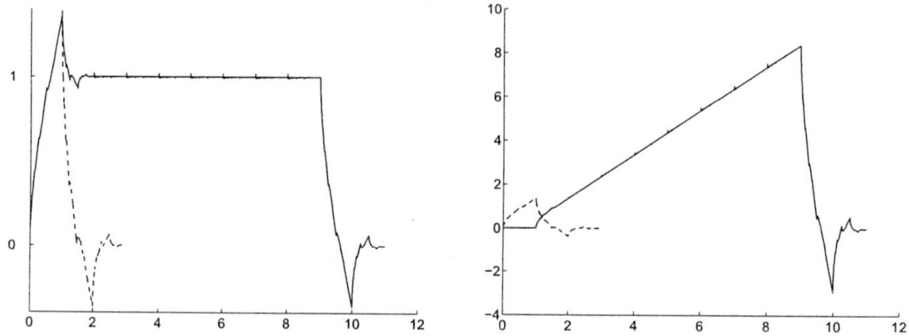

Figure 4.3: Db2 wavelet representation of constant and linear function

Figure represents approximations of the constant (left) and linear function (right) by the sum $\sum_{k=0}^{8} \varphi(x-k)$. The scaling function is Daubechies Db2 function and it's graph is drawn by a dash line. The compact support of the sum is the interval $[0, 11]$, but on the first two and the last two unit intervals the graph is false because all translations of the scaling function, which include these intervals into their supports, are not used. Functions $\varphi(x+2)$, $\varphi(x+1)$, $\varphi(x-9)$ and $\varphi(x-10)$ miss. Both approximations are accurate on the interval $[2, 9]$, because all translations of the scaling function that are different from zero on unit subintervals of this interval are included into the sum.

The parameter r in the condition (22) is at least equal to one. When the decomposition (25) is written for the left eigenvector $\mathbf{e}_0^\top$, having all of its components equal to one, we arrive at the condition (21). The inner product of the identity (21) with a wavelet $\psi(x)$, considering the scaling function and wavelet orthogonality, brings us to the well know conclusion that a wavelet is orthogonal to the unit, $\int \psi(x)\, dx = 0$. It means that the mean value, i.e. the first moment of the wavelet equals zero.

THEOREM 10. *If the* CONDITION A_r *(22) is fulfilled, the first* r *moments of the wavelet* $\psi(x)$ *are equal to zero,*

$$\int_{-\infty}^{\infty} x^m\, \psi(x)\, dx = 0, \qquad m = 0, 1, \ldots, r-1. \tag{26}$$

Proof: Inner products of identities (25) by a wavelet $\psi(x)$, due to the wavelet and scaling function orthogonality, follow that

$$\begin{aligned}\int_{-\infty}^{\infty} x^m\, \psi(x)\, dx &= \int_{-\infty}^{\infty} \sum_k y_m(k)\, \varphi(x+k)\psi(x)\, dx \\ &= \sum_k y_m(k) \int_{-\infty}^{\infty} \varphi(x+k)\, \psi(x)\, dx = 0, \qquad m = 0, \ldots, r-1.\end{aligned}$$

∎

This wavelet property has the consequence that an algebraic function with a degree less than r can only be represented by a scaling function per the first sun in decomposition (24). The details equal zero because all wavelet coefficients are $b_{jk} = 0$.

EXAMPLE 8. The elementary polynomials 1 and x can be represented by translations of the Daubechies scaling function defined by $N = 4$ coefficients (3.37) (Figure 4.3). The matrix M_3 is, according to the Example 7, equal to

$$M_3 = \frac{1}{4}\begin{pmatrix} 1+\sqrt{3} & 0 & 0 \\ 3-\sqrt{3} & 3+\sqrt{3} & 1+\sqrt{3} \\ 0 & 1-\sqrt{3} & 3-\sqrt{3} \end{pmatrix}.$$

Only two eigenvalues of this matrix,

$$\lambda_0 = 1 \qquad \mathbf{y}_0^\top = (1 \quad 1 \quad 1)$$
$$\lambda_1 = \frac{1}{2} \qquad \mathbf{y}_1^\top = \frac{1}{2}\,(\sqrt{3}-3 \quad \sqrt{3}-1 \quad \sqrt{3}+1)$$
$$\lambda_2 = \frac{1+\sqrt{3}}{4} \qquad \mathbf{y}_2^\top = (1 \quad 0 \quad 0)\,,$$

are powers of two. Corresponding eigenvectors, the constant vector $(c,\,c,\,c)^\top$ attached to $\lambda_0 = 1$ and linear vector $(c,\,c+1,\,c+2)^\top$, $c = \text{const}$, attached to $\lambda_1 = 1/2$, reconstruct polynomials of the order 0 and 1. The function $\sum y_0(k)\,\varphi(x+k)$ equals constant and the function $\sum y_1(k)\,\varphi(x+k)$ equals a multiple of x. Thus the Daubechies space $\mathcal{V}_0$ contains the functions 1 and x on every finite interval. These functions are orthogonal to the wavelets in the space $\mathcal{W}_0$,

$$\int \psi(x)\,dt = 0, \qquad \int x\,\psi(x)\,dx = 0,$$

meaning that the Daubechies wavelet has two moments equal to zero (two vanishing moments). ∎

From Theorem 9 follows that the approximation is more accurate when parameter r in the CONDITION A_r is greater. If the function $f(x)$ is a polynomial of the order less then r, it can be locally exactly represented by the basis functions $\{\varphi(x-k)\}$. If the function $f(x)$ is not a polynomial but is sufficiently smooth, we can represent it approximately on each interval by it's Taylor polynomial of the degree $r-1$. The approximation error is defined by the first member of the Taylor decomposition which cannot be reconstructed, yielding the expression $(\Delta x)^r f^{(r)}$ in the error.

THEOREM 11. *If the* CONDITION A_r *(22) is fulfilled every function* $f(x)$, *which is* r *times differentiable, can be approximated with an error of the order* $(\Delta x)^r = 2^{jr}$ *by its projection* $f_j(x)$ *onto the space* $\mathcal{V}_j$,

$$\|f(x) - f_j(x)\| \le C\,(\Delta x)^r\,\|f^{(r)}\|,$$

i.e.

$$\|f(x) - \sum_k a_{j,k}\,2^{-j/2}\,\varphi(2^{-j}x-k)\| \le C\,2^{jr}\,\|f^{(r)}\|. \tag{27}$$

∎

The efficiency of the approximation (24) depends on the number of coefficients $a_{J,k}$ and $b_{j,k}$ in the representation. The estimate for the coefficients $b_{j,k}$ is also determined by the first member of the Taylor decomposition which we cannot reconstruct, because the first r addends are canceled out based on the property (26) stating that the moments equal zero. This conclusion is formulated in the next theorem.

THEOREM 12. *If the* CONDITION A_r *(22) is fulfilled and if the function* $f(x)$ *is* r *times differentiable, its wavelet coefficients decrease as* 2^{jr} *for* $j \to -\infty$,

$$|b_{jk}| = \left| \int f(x)\, 2^{-j/2}\, \psi(2^{-j}x - k)\, dx \right| \leq C\, 2^{jr}\, \|f^{(r)}\|. \tag{28}$$

∎

Wavelet coefficients b_{jk} are directly linked to the local properties of a function $f(x)$. The number of basis functions necessary for an approximation of the desired accuracy depends on the number r. The greater values of the function smoothness order and parameter r affect that the decomposition coefficients tend to zero faster and less addends are needed for an accurate approximation. The core problem is to find an adequate basis that yields a good approximation with fewer basis functions. If the function is globally smooth, the Fourier basis (sine functions) is usually convenient. If the function has a finite jump the Fourier coefficients do not decrease faster than $1/j$ and a wavelet basis is a better choice for approximating of a piecewise smooth function. The coefficients will decrease more slowly only around the discontinuity. The wavelet basis, which is local, can separate the smooth parts so that around the discontinuity a finer resolution on smaller scales is used. This is the essence advantage of a wavelet approximation.

For a box function and Haar wavelet is $r = 1$. The main issue is choosing the adequate parameter r in practice. The condition of having the function $\varphi(x)$ be twice differentiable and the choice $r \approx 4$ is usually satisfactory. The smoothness of the scaling function depends on the convergence condition of the cascade algorithm, considering that in a general case this function is not known in an analytic form but is determined by the limit value of the cascade algorithm (if it exists).

4.5 Convergence

All properties analyzed in the previous section depend on the eigenvalues of the matrix M, defined by formula (13). The existence of the dilatation equation solution and its smoothness depend on the properties of the matrix T, defined by formula (14). If CONDITION A_r (22) is met numbers $1, 1/2, \ldots, 1/2^{2r-1}$ are the first $2r$ eigenvalues of the matrix T. The cascade algorithm convergence and the smoothness of the approximation depend on the other eigenvalues of this matrix.

Let us see first when the cascade algorithm (2), that can be written in the following form

$$\varphi^{(j+1)}(x) = \sum_k 2\, h(k)\, \varphi^{(j)}(2x - k), \quad j = 0, 1, \ldots, \tag{29}$$

converges. The standard choice for the initial approximation $\varphi^{(0)}(x)$ is the box function (Example 3.2). Some other function may also be chosen as the initial approximation.

LEMMA 1. *A necessary condition for the convergence of the cascade algorithm (29), where coefficients $h(k)$ satisfy the condition (20), is that the initial approximation $\varphi^{(0)}(x)$ satisfies the identity (21),*

$$\sum_k \varphi^{(0)}(x-k) \equiv 1.$$

Proof: Let us prove that the identity prevents oscillations of the scaling function sum on each iteration step,

$$P^{(j)}(x) = \sum_k \varphi^{(j)}(x-k) \qquad j = 0, 1, \ldots.$$

Really, from the first step of the cascade algorithm and the condition (20) we have

$$P^{(1)}(x) = \sum_n \varphi^{(1)}(x-n) = \sum_n \left(2 \sum_k h(k)\, \varphi^{(0)}(2(x-n)-k) \right)$$

$$= \sum_l \varphi^{(0)}(2x-l) \left(2 \sum_n h(l-2n) \right) = P^{(0)}(2x).$$

It follows by induction that

$$P^{(j)}(x) = P^{(0)}(2^j\, x),$$

and to prevent oscillations condition $P^{(0)}(x) \equiv 1$ has to be fulfilled. The box function obviously satisfies this condition. ∎

Next example shows what happens when the initial condition is not fulfilled.

EXAMPLE 9. It is clear that the box function is the fix point of the cascade algorithm

$$\varphi^{(j+1)}(x) = \varphi^{(j)}(2x) + \varphi^{(j)}(2x-1), \qquad j = 0, 1, \ldots.$$

This iterative process converges when we choose for the initial function $\varphi^{(0)}(x)$ the linear B-spline with the support $[0, 2]$ (Figure 4.4, (a)), and does not converge for the initial choice of the linear B-spline with the support $[0, 1]$ (Figure 4.4, (b)). In the last case $\varphi^{(j)}(x)$ has the saw shape and does not converge to the box function when $j \to \infty$. The initial function (b) does not satisfy the condition (21) while the initial function (a) satisfies it. ∎

In the above analysis we set only the basic requirement per coefficients $h(k)$. The convergence depends much more on the equation coefficients that specify the properties of the matrix T, since the $\mathcal{L}_2$ convergence of the cascade algorithm boils down to the convergence of the step iterative method defined by the sequence of vectors $\mathbf{a}^{(j)} = \{a^{(j)}(n)\}_n, \quad j = 0, 1, \ldots,$

(a)

$$\varphi^{(0)}(x) = \begin{cases} x, & x \in [0,1) \\ 2-x, & x \in [1,2) \end{cases}$$

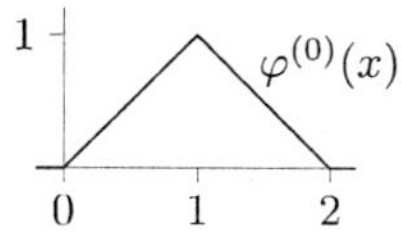

$$\varphi^{(0)}(2x) = \begin{cases} 2x, & x \in [0,\frac{1}{2}) \\ 2-2x, & x \in [\frac{1}{2},1) \end{cases}$$

$$\varphi^{(0)}(2x-1) = \begin{cases} 2x-1, & x \in [\frac{1}{2},1) \\ 3-2x, & x \in [1,\frac{3}{2}) \end{cases}$$

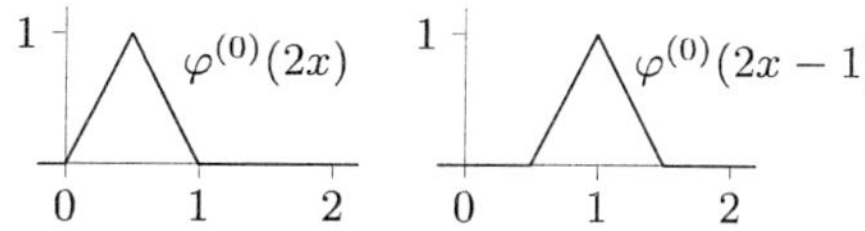

$$\varphi^{(1)}(x) = \begin{cases} 2x, & x \in [0,\frac{1}{2}) \\ 1, & x \in [\frac{1}{2},1) \\ 3-2x, & x \in [1,\frac{3}{2}) \end{cases}$$

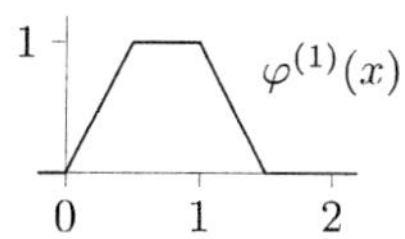

$$\varphi^{(j)}(x) = \begin{cases} 2^j x, & x \in [0,1/2^j) \\ 1, & x \in [1/2^j,1) \\ 2^j+1-2^j x, & x \in [1,(2^j+1)/2^j) \end{cases}$$

(b)

$$\varphi^{(0)}(x) = \begin{cases} 2x, & t \in [0,\frac{1}{2}) \\ 2-2x, & x \in [\frac{1}{2},1) \end{cases}$$

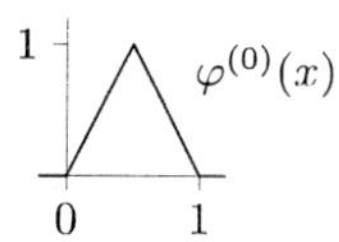

$$\varphi^{(0)}(2x) = \begin{cases} 4x, & x \in [0,\frac{1}{4}) \\ 2-4x, & x \in [\frac{1}{4},\frac{1}{2}) \end{cases}$$

$$\varphi^{(0)}(2x-1) = \begin{cases} 4x-2, & x \in [\frac{1}{2},\frac{3}{4}) \\ 4-4x, & x \in [\frac{3}{4},1) \end{cases}$$

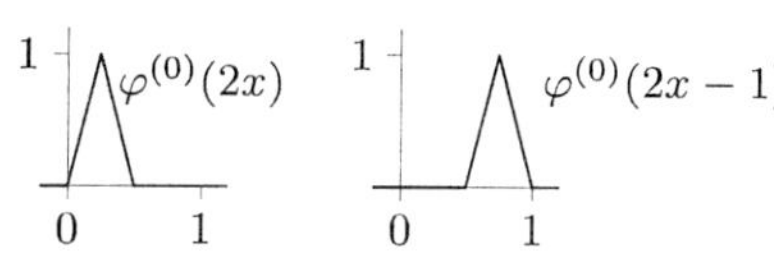

$$\varphi^{(1)}(x) = \begin{cases} 4x, & x \in [0,\frac{1}{4}) \\ 2-4x, & x \in [\frac{1}{4},\frac{1}{2}) \\ 4x-2, & x \in [\frac{1}{2},\frac{3}{4}) \\ 4-4x, & x \in [\frac{3}{4},1) \end{cases}$$

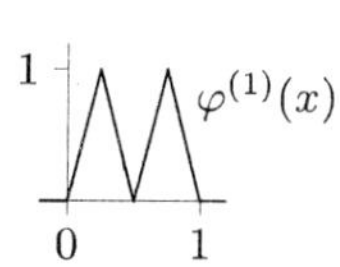

$$\varphi^{(j)}(x) = \begin{cases} 0, & x = (2k)/2^{j+1} \\ 1, & x = (2k+1)/2^{j+1} \end{cases}, \quad k = 0,\ldots,2^j-1.$$

Figure 4.4: Effect of the initial function to the cascade algorithm convergence

(30)
$$\mathbf{a}^{(j+1)} = T\,\mathbf{a}^{(j)}, \qquad j = 0, 1, \ldots,$$
$$a^{(j)}(n) = \big(\varphi^{(j)}(x),\, \varphi^{(j)}(x+n)\big) = \int \varphi^{(j)}(x)\,\varphi^{(j)}(x+n)\,dx,$$

for the equation

(31)
$$\mathbf{a} = T\,\mathbf{a}, \qquad \mathbf{a} = \{a(n)\}, \quad a(n) = \big(\varphi(x),\, \varphi(x+n)\big).$$

From Theorem 6 follows that the vector $\mathbf{e}^\top = (\ldots \;\; 1 \;\; \ldots \;\; 1 \;\; \ldots)$ is the left eigenvector of the matrix T,

$$\mathbf{e}^\top T = \mathbf{e}^\top M\,F^\top = \mathbf{e}^\top F^\top = \mathbf{e}^\top,$$

and the corresponding eigenvalue is $\lambda = 1$. It means that the sum of elements by matrix column equals one.

Let us illustrate the dependency of the iterative algorithm convergence on the eigenvalues of the matrix T by some examples.

EXAMPLE 10. Let us choose the coefficients $h(0) = 1$, $h(1) = 1/2$ and $h(2) = -1/2$ for which the sums of the coefficients with even and odd indices equal 1/2 (formula (20)). The dilatation equation (3) has the form

$$\varphi(x) = 2\,\varphi(2x) + \varphi(2x-1) - \varphi(2x-2).$$

At the point $x = 0$ the value of the scaling function doubles at each step of the cascade algorithm, because $\varphi^{(j+1)}(0) = 2\varphi^{(j)}(0)$, meaning that the cascade algorithm diverges whatever initial function is.

In order to analyze the convergence by the matrix T, let us first calculate

$$2F\,F^\top = \frac{1}{2}\begin{pmatrix} \cdot & & & & \\ 1 & 2 & & & \\ -1 & 1 & 2 & & \\ & -1 & 1 & 2 & \end{pmatrix}\begin{pmatrix} \cdot & 1 & -1 & \\ & 2 & 1 & -1 \\ & & 2 & 1 \\ & & & 2 \end{pmatrix} = \frac{1}{2}\begin{pmatrix} \ddots & \ddots & \ddots & \ddots & \\ -2 & 1 & 6 & 1 & -2 \\ & \ddots & \ddots & \ddots & \ddots \end{pmatrix}.$$

The operator $(\downarrow 2)$ performs a double shift, i.e. the odd rows are removed, thus

$$T = (\downarrow 2)\,2\,F\,F^\top = \frac{1}{2}\begin{pmatrix} \cdot & & & & & & \\ \cdot & 6 & 1 & -2 & & & \\ & -2 & 1 & 6 & 1 & -2 & \\ & & & -2 & 1 & 6 & \\ & & & & & -2 & \cdot \\ & & & & & & \cdot \end{pmatrix}$$

In all columns of the matrix T the sum of the elements equals one. As there are three dilatation coefficients, based on Theorem 1 for $N = 3$, the compact support of the scaling function is interval $[0, 2]$. Inner products defined in (30) equal zero for $|n| \geq 2$ and the iterative algorithm is determined only by the central sub-matrix T_3 of the order $2(N-1) - 1 = 3$,

$$T_3 = \frac{1}{2}\begin{pmatrix} 1 & -2 & 0 \\ 1 & 6 & 1 \\ 0 & -2 & 1 \end{pmatrix}.$$

If the initial function of the cascade algorithm is the box function the initial vector is $\mathbf{a}^{(0)} = (\ldots, 0, 1, 0, \ldots)^\top$ and we get the sequence

$$\mathbf{a}_3^{(0)} = \begin{pmatrix} 0 \\ 1 \\ 0 \end{pmatrix}, \qquad \mathbf{a}_3^{(1)} = T_3\,\mathbf{a}_3^{(0)} = \begin{pmatrix} -1 \\ 3 \\ -1 \end{pmatrix}, \qquad \mathbf{a}_3^{(2)} = T_3\,\mathbf{a}_3^{(1)} = \frac{1}{2}\begin{pmatrix} -7 \\ 16 \\ -7 \end{pmatrix}, \quad \cdots$$

It is obvious that numbers increase, i.e. that the cascade algorithm diverges. The matrix T_3 has the eigenvalue $\lambda = 5/2$, which is greater than one. ∎

EXAMPLE 11. The coefficient vector $\mathbf{h}$ with nonzero elements $h(0) = 1/2$ and $h(3) = 1/2$ meet the condition (20) and is orthogonal to the double shift,

$$\sum_k h(k)\,h(k-2m) = \frac{1}{4}\,\delta(m), \qquad m \in \mathcal{Z}.$$

The solution of the dilatation equation

$$\varphi(x) = \varphi(2x) + \varphi(2x-3), \qquad \text{is} \qquad \varphi_{[0,3]}(x) = \begin{cases} 1/3, & 0 \leq x < 3 \\ 0 & \text{otherwise} \end{cases}.$$

However, the cascade algorithm only weakly converges to the function $\varphi_{[0,3]}(x)$.

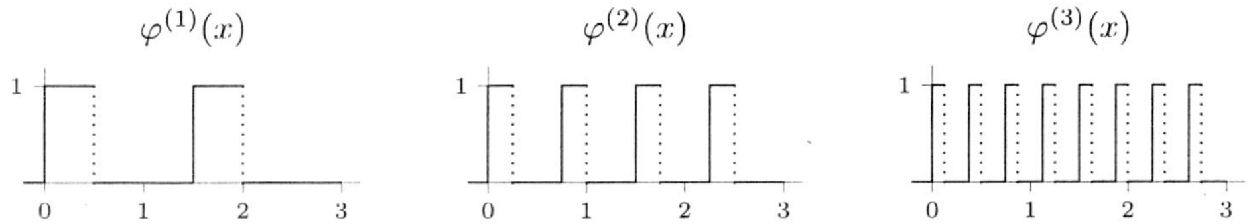

Figure 4.5: Weak convergence of the cascade algorithm

According to Figure 4.5, functions $\varphi^{(j)}(x)$ are not equal to $1/3$ in any of the points, but are equal to 0 or 1. However, the surface determined by these functions on the interval $[0,\,3]$ is equal to $1/3$. Fast oscillations of the functions $\varphi^{(j)}(x)$ are averaged by integration, thus we are dealing with a weak convergence of these functions towards the solution of the dilatation equation $\varphi_{[0,3]}(x)$. The attribute "weak" means that for every smooth function $f(x)$ the inner product converges,

$$\left(\varphi^{(j)}(x),\, f(x)\right) \quad \underset{j\to\infty}{\longrightarrow} \quad \left(\varphi_{[0,3]}(x),\, f(x)\right),$$

i.e.

$$\lim_{j\to\infty} \int_0^3 \varphi^{(j)}(x)\, f(x)\, dx = \int_0^3 \frac{1}{3}\, f(x)\, dx.$$

Iterations (30) are defined by the central sub-matrix with the dimension 5, since $N = 4$ (the last nonzero coefficient is $h(3)$),

$$T_5 = \frac{1}{2} \begin{pmatrix} 0 & 1 & 0 & 0 & 0 \\ 2 & 0 & 0 & 1 & 0 \\ 0 & 0 & 2 & 0 & 0 \\ 0 & 1 & 0 & 0 & 2 \\ 0 & 0 & 0 & 1 & 0 \end{pmatrix}.$$

The sum of elements in each column in this matrix equals one. The matrix T_5 has eigenvalues -1, $-\frac{1}{2}$ and $\frac{1}{2}$ and the double eigenvalue $\lambda = 1$. The cascade algorithm does not converge in every point, but only in the mean towards the dilated box function.

The right eigenvectors that correspond to the double eigenvalue $\lambda = 1$ are

$$\mathbf{t}_1 = \begin{pmatrix} 0 & 0 & 1 & 0 & 0 \end{pmatrix}^\top, \qquad \mathbf{t}_2 = \frac{1}{9}\begin{pmatrix} 1 & 2 & 3 & 2 & 1 \end{pmatrix}^\top,$$

and they are both solutions of the eigenvalue problem (31). If we choose the box function as the initial function of the cascade algorithm, the inner products of the box function and its translations determine the initial vector $\mathbf{a}_5^{(0)} = \mathbf{t}_1$. Multiplication by T_5 yields the same vector $\mathbf{a}_5^{(1)} = \mathbf{a}_5^{(0)}$. In general, at each step

$$\mathbf{a}_5^{(j)} = \mathbf{t}_1 = \begin{pmatrix} 0 & 0 & 1 & 0 & 0 \end{pmatrix}^\top,$$

which means that the function $\varphi^{(j)}(x)$ is orthogonal to its translations for arbitrary j. However, the inner products of the dilated box function $\varphi_{[0,3]}(x)$ in the interval $0 \le x < 3$ and its translations are elements of the second eigenvector $\mathbf{t}_2$. The function $\varphi_{[0,3]}(x)$ is not orthogonal to its translations. A consequence of the weak convergence is that the dilated box function is not orthogonal to its translations, whereas the functions $\varphi^{(j)}(x)$ are,

$$a^{(j)}(n) = \left(\varphi^{(j)}(x),\, \varphi^{(j)}(x+n)\right) = \delta(n) \quad \not\rightarrow \quad a(n) = \left(\varphi_{[0,3]}(x),\, \varphi_{[0,3]}(x+n)\right).$$

At each step $\varphi^{(j)}(x)$ has the unit energy, while the energy of the limiting function is $\int \left(\varphi(x)\right)^2 dx = 1/3$.

∎

EXAMPLE 12. The coefficients $h(0) = 1/4$, $h(1) = 1/2$ and $h(2) = 1/4$ define the dilatation equation for the roof function. It is shown that the cascade algorithm converges to this function in Example 8. The matrix $2FF^\top$ has the elements $\frac{1}{8}(1, 4, 6, 4, 1)$. Since the number of the nonzero coefficients is $N = 3$, the transformation matrix is

$$T_3 = \frac{1}{8}\begin{pmatrix} 4 & 1 & 0 \\ 4 & 6 & 4 \\ 0 & 1 & 4 \end{pmatrix}.$$

with the eigenvalues of 1, 1/2 and 1/4. Starting from the initial approximation determined by the box function $\mathbf{a}_3^{(0)} = \begin{pmatrix} 0 & 1 & 0 \end{pmatrix}^\top$, by multiplying with T_3 we arrive, in the following steps of the cascade algorithm, to the vectors

$$\mathbf{a}_3^{(1)} = \frac{1}{8}\begin{pmatrix} 1 \\ 6 \\ 1 \end{pmatrix} \qquad \mathbf{a}_3^{(2)} = \frac{1}{64}\begin{pmatrix} 10 \\ 44 \\ 10 \end{pmatrix} \qquad \cdots \qquad \longrightarrow \qquad \mathbf{a}_3 = \frac{1}{6}\begin{pmatrix} 1 \\ 4 \\ 1 \end{pmatrix}.$$

$\mathbf{a}_3$ is the eigenvector of the matrix T_3 corresponding to the eigenvalue of $\lambda = 1$, with elements that are the inner products of the roof function and its translations. The sequence of functions $\varphi^{(j)}(x)$ uniformly (in every point) converges towards the roof function. ∎

Note that the cascade algorithm in Example 10 diverges, while the matrix T has the eigenvalue $|\lambda| > 1$. In Example 11 we only have a weak convergence of the algorithm, while the matrix T has a double eigenvalue $\lambda = 1$, or several eigenvalues with the module equal to one. In Example 12 the cascade algorithm uniformly converges; $\lambda = 1$ is the unique eigenvalue, while all other eigenvalues satisfy relation $|\lambda| < 1$. The conditions for uniform convergence of the cascade algorithm are formulated by the following theorem.

THEOREM 13. *Let us assume that the scaling function $\varphi(x) \in \mathcal{L}_2$. The cascade sequence $\varphi^{(j)}(x)$ converges towards the scaling function $\varphi(x)$ in the $\mathcal{L}_2$ norm,*

$$\lim_{j\to\infty} \|\varphi^{(j)}(x) - \varphi(x)\|^2 = 0,$$

if and only if T has a single eigenvalue $\lambda = 1$, and all other eigenvalues satisfy condition $|\lambda| < 1$.

The proof can be find in [44]. ∎

The interpretation of the convergence of the cascade algorithm in the frequency domain boils down to the algorithm based on the Fourier transform (§4.2),

$$\hat{\varphi}^{(j)}(\omega) = \left(\prod_{k=1}^{j} \hat{h}(\omega/2^k)\right) \hat{\varphi}^{(0)}(\omega/2^j) \qquad \underset{j\to\infty}{\longrightarrow} \qquad \hat{\varphi}(\omega) = \prod_{k=1}^{\infty} \hat{h}(\omega/2^k).$$

In accordance with what has been said above, if the eigenvalues of the matrix T are less than one, except for a single one equalling one, the limiting function of the cascade algorithm $\varphi(x)$ belongs to the space $\mathcal{L}_2$. It represents the minimum smoothness. Following two theorems more precisely define smoothness conditions for the scaling function and wavelets.

THEOREM 14. *If the eigenvalues of the matrix T, which are not powers of $1/2$, meet the condition $|\lambda| < 4^{-s}$, $\varphi(x)$ and $\psi(x)$ have s derivatives in $\mathcal{L}_2$. The upper limit s_{max} of the numbers s, when $\varphi(x)$ is determined by the frequency response (8) of the form*

$$H(z) = \left(\frac{1+z^{-1}}{2}\right)^r Q(z),$$

equals

$$s_{max} = r - \log_4|\lambda_{max}(T_Q)|, \qquad T_Q = (\downarrow 2)\, 2\, F_Q\, F_Q^{\top}. \tag{32}$$

The proof can be found in [44]. ∎

THEOREM 15. *If the scaling function $\varphi(x)$ has s derivatives in $\mathcal{L}_2$ and r is the order of approximation error the estimate $s < r$ stands. If we allow for s not being an integer, it is necessary that $s \le r - 1/2$. Uniformly (in all points) the smoothness order cannot be greater than $r - 1$.*

Proof: : If the function $\varphi(x) \in \mathcal{L}_2$ the frequency response $\hat{h}(\omega)$ needs to have at least one zero in point π (CONDITION A_1 given by (22) for $r = 1$). Analogue to that, in order to have $\varphi^{(s)}(x) \in \mathcal{L}_2$, the frequency response needs to have a zero of at least the order $(s+1)$ in point π. Thus $r \ge s+1$. ∎

5

How to compute

In wavelet analysis we usually talk about approximations and details. Approximations are low-frequency components of a function at large scales represented by the first addend, while details are high-frequency components of a function at smaller scales represented by the second addend in formula (4.24). The wavelet transform of a function has as an output scaling function coefficients $a_{J,k}$ (approximation) and wavelet coefficients $b_{j,k}$ (details). A scaling function and a wavelet are defined in an analytic form very rare, they are mainly defined by recursion. The recursive nature of the dilatation equation characterizes the wavelet transform (§2.4). Integrals defining the inner products are not calculated, instead the coefficients in the representation of a function on one multiresolution level are determined by the same coefficients on the previous multiresolution level and the coefficients of the dilatation equation. The calculation procedure is extremely fast and efficient in the case of the orthnormal basis. It is based on the pyramid algorithm, already presented in §3.3.

5.1 Discrete wavelet transform

Discrete Wavelet Transform (DWT) is an algorithm for determining the wavelet and scaling function coefficients at the dyadic scales and in the dyadic points.

One step in the process of analysis consists of the separation of the approximation and details of the discrete signal, thus yielding two signals as a result. Both signals are the same length as the initial, doubling the amount of data. Compression, or the discarding of every other piece of information, halves the length of the output signals, so that the total amount of data at the output equals the amount of data at the input. The approximation arrived at is the input signal for the next step (Figure 5.1).

The effect of the transformations performed is a coarser time, but finer frequency resolution of the output signals. The compression halves the time resolution, because now only half of the total number of samples characterizes the entire sig-

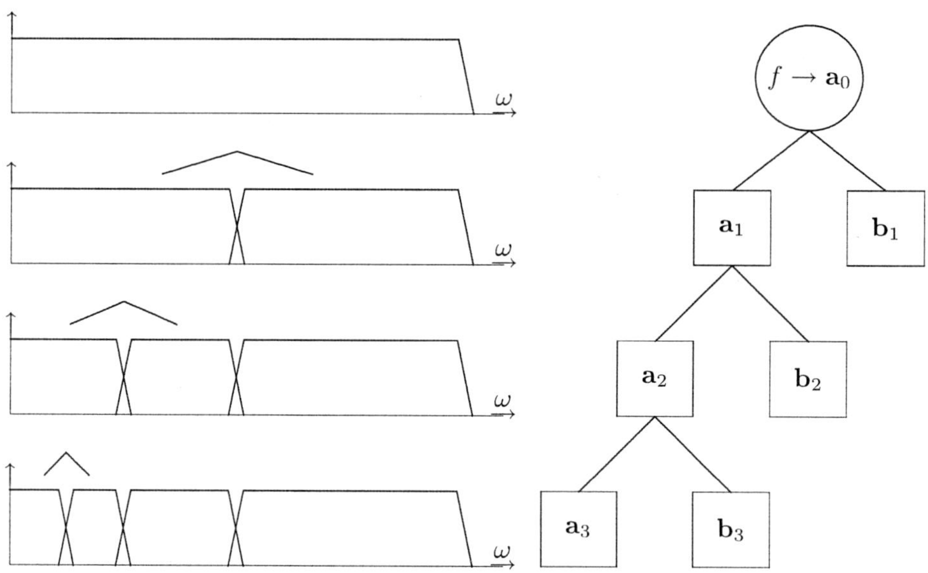

Figure 5.1: Discrete Wavelet Transform (DWT)

nal. However, the decomposition doubles the frequency resolution, because the frequency layer of each of the output signals includes only half of the previous frequency layer. The procedure described is the sub-band coding in signal processing, and can be repeated for further decomposition. At each level filtering and compression will halve the frequency layer thus doubling the frequency resolution, and reduce the number of samples by half thus doubling the time step and halving the time resolution. Finally, if the original signal is 2^m in length DWT has at most m steps and at the output the approximation is a signal of one in length. The DWT of the original signal is obtained by connecting all the coefficients starting from the last decomposition level. It is a vector made up of the output signals $[\mathbf{a}_J, \mathbf{b}_J, \ldots, \mathbf{b}_2, \mathbf{b}_1]$. The total number of the DWT coefficients equals the length of the initial signal.

The algorithm described, representing the essence of the discrete wavelet transform, is used for the analysis, i.e. decomposition of signals. Assembling the components in order to gain the initial signal with no loss of information is called reconstruction or synthesis. The mathematical operations used to perform synthesis are called *Inverse Discrete Wavelet Transform* (IDWT). The wavelet analysis includes filtering and compression, while the wavelet reconstruction process is made up of decompression and filtering. It is necessary to reconstruct the approximation and details, before they are combined. Shortly,

- The analysis DWT algorithm starts with the function f that defines the initial approximation $\mathbf{a}_0$ (see paragraph *Initial choice of coefficients* below). Then signals $\mathbf{a}_1$ and $\mathbf{b}_1$ are determined from $\mathbf{a}_0$, next $\mathbf{a}_2$ and $\mathbf{b}_2$ are determined from $\mathbf{a}_1$, etc.

- The synthesis IDWT algorithm starts from the signals $\mathbf{a}_J$ and $\mathbf{b}_J$, then the signal $\mathbf{a}_{J-1}$ is calculated based on them, next using the signals $\mathbf{a}_{J-1}$ and $\mathbf{b}_{J-1}$ the signal $\mathbf{a}_{J-2}$ is determined, etc.

Fast Wavelet Transform (FWT). In 1988, Mallat [30] developed an efficient procedure for the decomposition of signals into approximation and details using the matrix interpretation of the pyramid algorithm (§3.3).

THEOREM 1. *For the function* $\sum_l a_{j-1,l}\varphi_{j-1,l}(x)$ *belonging to the space* $\mathcal{V}_{j-1} = \mathcal{V}_j \oplus \mathcal{W}_j$, *the Fourier coefficients* $\mathbf{a}_j = \{a_{j,k}\}$ *and* $\mathbf{b}_j = \{b_{j,k}\}$ *by the new orthonormal basis* $\{\varphi_{j,k}(x),\, \psi_{j,k}(x)\}$ *are calculated through the recursion*

$$\mathbf{a}_j = (\downarrow 2)\, C^\top \mathbf{a}_{j-1}, \qquad \mathbf{b}_j = (\downarrow 2)\, D^\top \mathbf{a}_{j-1}, \tag{1}$$

which can be represented using the diagram

$$\begin{array}{ccccccccc}
\mathbf{a}_0 & \xrightarrow{C^\top} & \mathbf{a}_1 & \xrightarrow{C^\top} & \mathbf{a}_2 & \longrightarrow & \cdots\ \mathbf{a}_{J-1} & \xrightarrow{C^\top} & \mathbf{a}_J \\
& D^\top \searrow & & D^\top \searrow & & \searrow & & D^\top \searrow & \\
& & \mathbf{b}_1 & & \mathbf{b}_2 & & & & \mathbf{b}_J
\end{array}$$

The filter matrices $C = \{c(k)\}$ *and* $D = \{d(k)\}$ *have the form (2.31) and are generated by the coefficients* $c(k)$ *of the dilatation equation (3.11) and the coefficients* $d(k)$ *of the wavelet equation (3.12).*

Proof: The statement is a matrix interpretation of Theorem 3.2. ∎

The filter matrices $C^\top$ and $D^\top$ are compressed by erasing every other row, thus yielding the rectangular matrices $(\downarrow 2)\, C^\top$ and $(\downarrow 2)\, D^\top$ with the number of columns being twice the number of rows. By merging these two matrices in a single square matrix by continuing the matrix $(\downarrow 2)\, D^\top$ below the matrix $(\downarrow 2)\, C^\top$, the discrete wavelet transform is performed in a single step through the multiplication of the input vector $\mathbf{a}_{j-1}$ by the matrix thus produced

$$\begin{pmatrix} \mathbf{a}_j \\ \mathbf{b}_j \end{pmatrix} = (\downarrow 2) \begin{pmatrix} C^\top \\ D^\top \end{pmatrix} \mathbf{a}_{j-1}. \tag{2}$$

During the process of reconstruction from the basis $\{\varphi_{j,k}(x),\, \psi_{j,k}(x)\}$ it is necessary to return to the basis $\{\varphi_{j-1,l}(x)\}$. Since the bases are orthonormal the filter matrix is orthogonal and the inverse matrix is equal to the transposed.

THEOREM 2. *The Fourier coefficients* $\mathbf{a}_{j-1} = \{a_{j-1,l}\}$ *of a function by the orthonormal basis* $\{\varphi_{j-1,l}(x)\}$ *are determined by the Fourier coefficients* $\mathbf{a}_j = \{a_{j,k}\}$ *and* $\mathbf{b}_j = \{b_{j,k}\}$ *by the orthonormal basis* $\{\varphi_{j,l}(x),\, \psi_{j,l}(x)\}$ *through the recursion*

$$\mathbf{a}_{j-1} = \begin{pmatrix} C & D \end{pmatrix} (\uparrow 2) \begin{pmatrix} \mathbf{a}_j \\ \mathbf{b}_j \end{pmatrix}, \tag{3}$$

and can be represented by the following diagram

$$\begin{array}{ccccccccc} \mathbf{a}_J & \overset{C}{\longrightarrow} & \mathbf{a}_{J-1} & \overset{C}{\longrightarrow} & \mathbf{a}_{J-2} & \cdots & \mathbf{a}_1 & \overset{C}{\longrightarrow} & \mathbf{a}_0 \\ & \nearrow D & & \nearrow D & & & & \nearrow D & \\ \mathbf{b}_J & & \mathbf{b}_{J-1} & & & & \mathbf{b}_1 & & \end{array}$$

Proof: The statement is a matrix interpretation of Theorem 3.3. The inverse fast wavelet transform is defined by inverting the matrix equation given in (2), which arrives at formula (3). ∎

It is clear that the discrete wavelet transform can be represented by multiplying a vector with a wavelet matrix W. If the transformation is performed by orthogonal wavelets the matrix W is orthogonal $W^{-1} = W^\top$. The wavelet transform $W\,M\,W^\top$ of an arbitrary matrix M is then the unitary, numerically stable transformation. The FWT algorithm is analogue to the FFT algorithm in Fourier analysis. For a signal N in length the calculation complexity is $O(N)$ in the case of FWT, while it is $O(N \log_2 N)$ in the case of FFT. The algorithm is fully recursive.

EXAMPLE 1. Let us illustrate the application of the FWT algorithm for the analysis of the signal $\mathbf{x} = (37, 35, 28, 28, 58, 18, 21, 15)$ with a dimension of $8 = 2^3$. The maximum number of decomposition levels in this case is $J = 3$. The initial sample is marked as level zero. For simplicity, we shall use the Haar wavelet determined by a box function (§3.4), orthogonal to the translation.

Level 0: We shall take the given signal as the initial approximation,

$$\mathbf{a}_0 = \begin{pmatrix} 37 & 35 & 28 & 28 & 58 & 18 & 21 & 15 \end{pmatrix}^\top$$

Level 1: According to (2) for $j = 1$, after the first step of FWT decomposition

$$\sqrt{2}\begin{pmatrix} \boxed{36} \\ \boxed{28} \\ \boxed{38} \\ \boxed{18} \\ 1 \\ 0 \\ 20 \\ 3 \end{pmatrix} = \frac{1}{\sqrt{2}} \begin{pmatrix} 1 & 1 & 0 & 0 & 0 & 0 & 0 & 0 \\ 0 & 0 & 1 & 1 & 0 & 0 & 0 & 0 \\ 0 & 0 & 0 & 0 & 1 & 1 & 0 & 0 \\ 0 & 0 & 0 & 0 & 0 & 0 & 1 & 1 \\ 1 & -1 & 0 & 0 & 0 & 0 & 0 & 0 \\ 0 & 0 & 1 & -1 & 0 & 0 & 0 & 0 \\ 0 & 0 & 0 & 0 & 1 & -1 & 0 & 0 \\ 0 & 0 & 0 & 0 & 0 & 0 & 1 & -1 \end{pmatrix} \begin{pmatrix} 37 \\ 35 \\ 28 \\ 28 \\ 58 \\ 18 \\ 21 \\ 15 \end{pmatrix}$$

we arrive at the approximation $\mathbf{a}_1 = \sqrt{2}\begin{pmatrix}36\\28\\38\\18\end{pmatrix}$ and the detail $\mathbf{b}_1 = \sqrt{2}\begin{pmatrix}1\\0\\20\\3\end{pmatrix}$,

because

$$C = \frac{1}{\sqrt{2}}\begin{pmatrix}1&0&0&0&0&0&0&0\\1&1&0&0&0&0&0&0\\0&1&1&0&0&0&0&0\\0&0&1&1&0&0&0&0\\0&0&0&1&1&0&0&0\\0&0&0&0&1&1&0&0\\0&0&0&0&0&1&1&0\\0&0&0&0&0&0&1&1\end{pmatrix}$$

$$(\downarrow 2)\,C^{\top} = \frac{1}{\sqrt{2}}\begin{pmatrix}1&1&0&0&0&0&0&0\\0&0&1&1&0&0&0&0\\0&0&0&0&1&1&0&0\\0&0&0&0&0&0&1&1\end{pmatrix}$$

$$D = \frac{1}{\sqrt{2}}\begin{pmatrix}1&0&0&0&0&0&0&0\\-1&1&0&0&0&0&0&0\\0&-1&1&0&0&0&0&0\\0&0&-1&1&0&0&0&0\\0&0&0&-1&1&0&0&0\\0&0&0&0&-1&1&0&0\\0&0&0&0&0&-1&1&0\\0&0&0&0&0&0&-1&1\end{pmatrix}$$

$$(\downarrow 2)\,D^{\top} = \frac{1}{\sqrt{2}}\begin{pmatrix}1&-1&0&0&0&0&0&0\\0&0&1&-1&0&0&0&0\\0&0&0&0&1&-1&0&0\\0&0&0&0&0&0&1&-1\end{pmatrix}$$

Level 2: The input signal is now an approximation determined by the vector $\mathbf{a}_1$ from the previous level,

$$2\begin{pmatrix}\boxed{32}\\\boxed{28}\\4\\10\end{pmatrix} = \frac{1}{\sqrt{2}}\begin{pmatrix}1&1&0&0\\0&0&1&1\\1&-1&0&0\\0&0&1&-1\end{pmatrix}\sqrt{2}\begin{pmatrix}36\\28\\38\\18\end{pmatrix},$$

yielding the approximation $\mathbf{a}_2 = 2\begin{pmatrix}32\\28\end{pmatrix}$ and the detail $\mathbf{b}_2 = 2\begin{pmatrix}4\\10\end{pmatrix}$.

Level 3: The application of the FWT algorithm to the approximation determined by the vector $\mathbf{a}_2$,

$$2\sqrt{2}\begin{pmatrix}\boxed{30}\\ 2\end{pmatrix} = \frac{1}{\sqrt{2}}\begin{pmatrix}1 & 1\\ 1 & -1\end{pmatrix} 2\begin{pmatrix}32\\ 28\end{pmatrix},$$

yields vectors with a dimension of one, the approximation $\mathbf{a}_3 = 2\sqrt{2}\,(30)$ and the detail $\mathbf{b}_3 = 2\sqrt{2}\,(2)$. It is the final possible level for this amount of data.

The full fast wavelet transform of the given signal can be briefly represented by the diagram:

$$\begin{array}{l}
\boxed{37\quad 35\quad 28\quad 28\quad 58\quad 18\quad 21\quad 15}\\
\qquad\qquad\swarrow \qquad\qquad\qquad \searrow\\
\boxed{36\quad 28\quad 38\quad 18}\qquad\qquad 1\quad 0\quad 20\quad 3\\
\qquad\swarrow \qquad \searrow\\
\boxed{32\quad 28}\qquad 4\quad 10\\
\swarrow \quad \searrow\\
\boxed{30}\qquad 2\\
\hline
\boxed{30}\qquad 2\qquad 4\quad 10\qquad 1\quad 0\quad 20\quad 3
\end{array} \tag{4}$$

The sequence of numbers under the line in diagram (4), determined by the approximation at the final level and details on all levels from the first to the last, represent the wavelet coefficients of the signal $\mathbf{x}$. We just have to explain why the factors $2^{j/2}$, appearing as multipliers in the vectors $\mathbf{a}_j$ and $\mathbf{b}_j$, have been left out of the scheme. This factor is reduced by the norming factor of the matching scaling function or wavelet, because, according to formula (4.24), the representation of the signal by the wavelets is

$$\begin{aligned}
x(n) &= \sum_k a_{J,k}\, 2^{-J/2}\, \varphi(2^{-J}n - k) + \sum_j \sum_k b_{j,k}\, 2^{-j/2}\, \psi(2^{-j}n - k)\\
&= \sum_k \bar{a}_{J,k}\, \varphi(2^{-J}n - k) + \sum_j \sum_k \bar{b}_{j,k}\, \psi(2^{-j}n - k).
\end{aligned}$$

The coefficients $\bar{a}_{J,k} = a_{J,k}\, 2^{-J/2}$ and $\bar{b}_{j,k} = b_{j,k}\, 2^{-j/2}$ are the very wavelet coefficients from scheme (4). The graphical display of the approximation and details of the signal $\mathbf{x}$ is given on Figure 5.2.

The reconstruction of the signal $\mathbf{x}$ based on the given vectors $\mathbf{a}_3$, $\mathbf{b}_3$, $\mathbf{b}_2$ and $\mathbf{b}_1$ is performed by the inverse pyramid algorithm with its matrix notation given by the expression (3). We will only write the final step of the reconstruction,

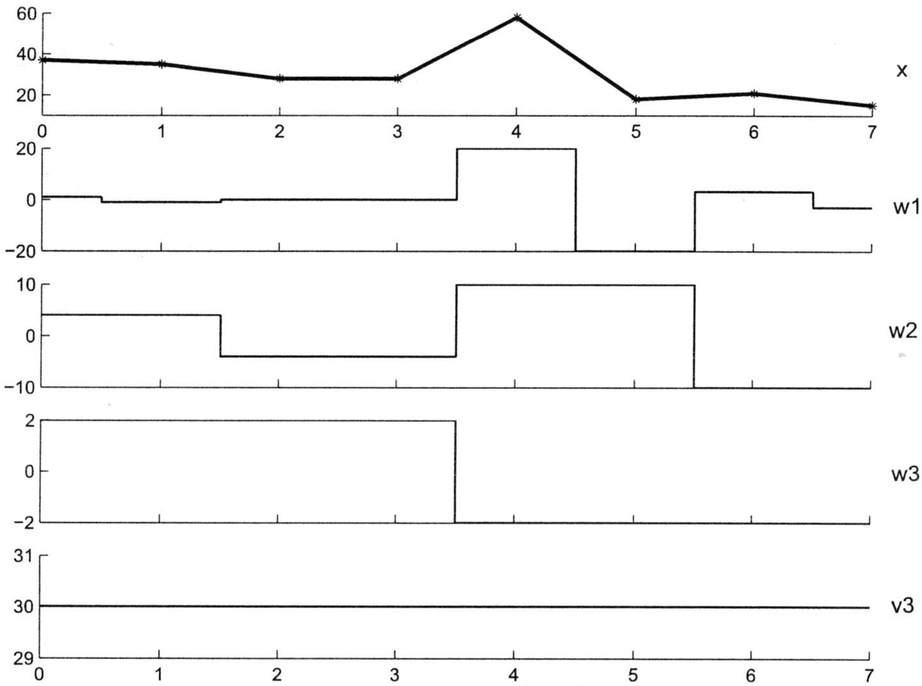

Figure 5.2: Signal components in approximation (v) and wavelet (w) spaces

considering that the algorithm is analogue to the decomposition algorithm described in details, except that in this case the transposed matrices are used,

$$\begin{pmatrix} 37 \\ 35 \\ 28 \\ 28 \\ 58 \\ 18 \\ 21 \\ 15 \end{pmatrix} = \frac{1}{\sqrt{2}} \begin{pmatrix} 1 & 0 & 0 & 0 & 1 & 0 & 0 & 0 \\ 1 & 0 & 0 & 0 & -1 & 0 & 0 & 0 \\ 0 & 1 & 0 & 0 & 0 & 1 & 0 & 0 \\ 0 & 1 & 0 & 0 & 0 & -1 & 0 & 0 \\ 0 & 0 & 1 & 0 & 0 & 0 & 1 & 0 \\ 0 & 0 & 1 & 0 & 0 & 0 & -1 & 0 \\ 0 & 0 & 0 & 1 & 0 & 0 & 0 & 1 \\ 0 & 0 & 0 & 1 & 0 & 0 & 0 & -1 \end{pmatrix} \cdot \sqrt{2} \begin{pmatrix} \boxed{36} \\ \boxed{28} \\ \boxed{38} \\ \boxed{18} \\ 1 \\ 0 \\ 20 \\ 3 \end{pmatrix}.$$

∎

For almost all signals the low-frequency content is the most important part. The high-frequency content determines details and nuances. Let us take as an example the human voice. If we remove the high-frequency components the voice is going to sound different, but we can still make out what is being said. However, if we remove enough low-frequency components the speech becomes unintelligible. An adequate choice of filters C and D in the discrete wavelet transform can produce very small by absolute value details $b_{j,k}$ at various levels j. In that case coding the signal is reduced to coding the final approximation $\mathbf{a}_J$ and those detail coefficients $b_{j,k}$ which are above the selected threshold. This operation is known as the compression of a signal. If a higher number of details can be neglected, considering the final approximation contains 2^J (J is the number of decomposition levels) times fewer data than the input signal, a considerable time and memory space saving is made in signal processing.

EXAMPLE 2. Let us illustrate the compression on the simple signal from Example 1, i.e. let us replace with zeroes all wavelet coefficients determined by the scheme (4) which are not greater than a given threshold. For two threshold values $TS = 2$ and $TS = 4$, the reconstruction of the modified signals is given in the following diagram:

$$\boxed{30} \quad 2 \quad 4 \quad 10 \quad 1 \quad 0 \quad 20 \quad 3$$

$\swarrow TS = 2$	$\searrow TS = 4$
[30] 0 4 10 0 0 20 3	[30] 0 0 10 0 0 20 0
[30 30] 4 10 0 0 20 3	[30 30] 0 10 0 0 20 0
[34 26 40 20] 0 0 20 3	[30 30 40 20] 0 0 20 0
[34 34 26 26 60 20 23 17]	[30 30 30 30 60 20 20 20]

With the threshold $TS = 2$ two of the wavelet coefficients are replaced by zero, and since one of the coefficients was already zero the signal is written with five, instead of eight pieces of data. With the threshold $TS = 4$ five wavelet coefficients become zero, thus the signal is only defined with three numbers. Figure 5.3 displays the "cutoff" effect for the two threshold values being analyzed. ∎

It is obvious that the signal is flattened more and more as the threshold is greater in this way. If a very precise reconstruction of the initial signal is not important while the compression is important we shall choose a higher threshold value. Thus the amount of information used to describe the signal is radically reduced on account of the quality of the reconstructed signal (in the sense of their proximity to the initial signal). The smaller the threshold, the less the output signal differs from the input, but more data is needed to define it.

Initial choice of coefficients. The issue of determining the initial sequence of coefficients $\{a_{0k}\}$ arises, as the vector $\mathbf{a}_0 = \{a_{0k}\}$ defines the initial step in the decomposition using the pyramid algorithm (1). These coefficients are Fourier coefficients of the function $f(x)$ by the orthonormal basis $\{\varphi(x-k)\}$,

$$a_{0,k} = \int f(x)\,\varphi(x-k)\,dx. \tag{5}$$

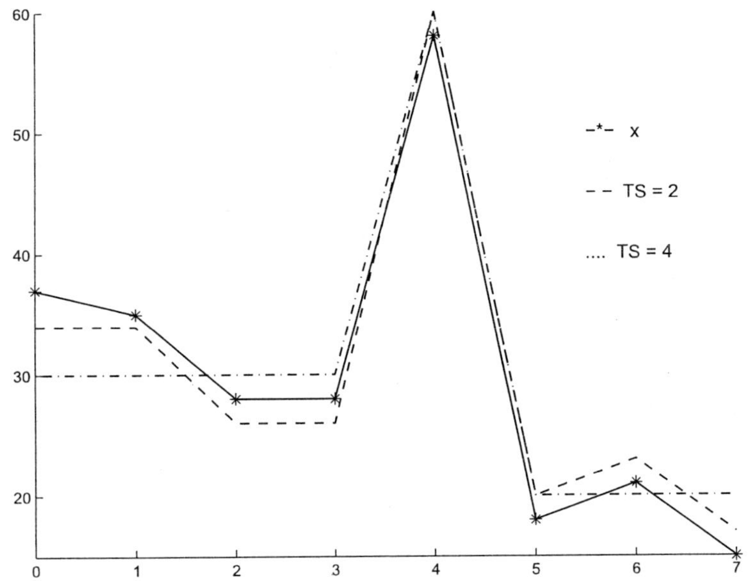

Figure 5.3: The initial and the compressed signals

If the function $f(x) \in \mathcal{V}_0$, it is accurately represented by the decomposition

$$f(x) = \sum_k a_{0,k}\, \varphi(x-k). \tag{6}$$

If $f(x) \notin \mathcal{V}_0$, the sum (6) represents an orthogonal projection of the function $f(x)$ onto the space $\mathcal{V}_0$. For integer argument values $x = n$ the sum (6) represents the discrete signal $\mathbf{f} = \{f(n)\}_n$,

$$f(n) = \sum_k a_{0,k}\, \varphi(n-k). \tag{7}$$

The coefficients sought in this case are the solution of the linear equation system (7). Finding the coefficients $a_{0,k}$ is called pre-filtering.

By approximating the integral in formula (5) by a sum pre-filtering is reduced to the approximate calculation of the coefficients sought by the finite sum

$$a_{0,k} \approx \sum_n f(n)\, \varphi(n-k). \tag{8}$$

In order to reconstruct the signal $\mathbf{f}$ during synthesis post-filtering needs to be performed (a procedure opposite to that described above) of the obtained coefficients $a_{0,k}$.

A scaling function and a wavelet graphs. The simplest way of generating the scaling function and wavelet graphs is based on the synthesis pyramid algorithm (3). If the cascade algorithm given by (4.2) converges, the limit function is the scaling function $\varphi(x)$. When we start from the box function $\varphi_b(x)$ as the initial function, the j-th iteration represents the approximation of the scaling function at a resolution of 2^{-j} by box functions. The approximation is a piecewise constant function at the dyadic intervals $\left[2^{-j}k, 2^{-j}(k+1)\right)$

$$\varphi^{(j)}(x) = \sum_k a_{-j,k}\,\varphi_b(2^j x - k), \qquad j = 0, 1, \ldots . \tag{9}$$

The coefficients in representation (9) are equal to

$$a_{-j,n} = \varphi^{(j)}(n\,2^{-j}) \quad \text{because} \quad \varphi_b\big(2^j(n\,2^{-j}) - k\big) = \delta(n-k).$$

Due to the convergence of the cascade algorithm, the values $a_{-j,n}$ tend to the exact values of the scaling function $\varphi(x)$ at the dyadic points. In approximations (9) for every j the wavelet coefficients $b_{-j,k}$ are equal to zero.

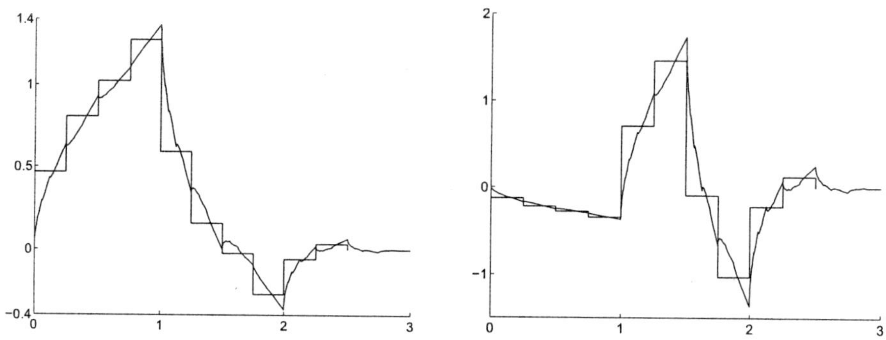

Figure 5.4: Approximations of Db2 scaling function and wavelet

Thus, in order to generate the scaling function we start from the coefficients $a_{0,n} = \delta(n)$ and $b_{0,n} = 0$ (we shifted indexes such that $J = 0$ in the inverse pyramid algorithm (3)). Taking into consideration the conclusion that $b_{-j,n} = 0$ for every j, the inverse pyramid algorithm (3) directly gives the approximations of the scaling function values at the dyadic points

$$\varphi\big(\frac{n}{2^{j+1}}\big) \approx a_{-(j+1),n} = \sum_k c(n-2k)\,a_{-j,k}, \qquad j = 0, 1, \ldots .$$

With an increase of the number of the resolution level (the iteration index j) the points become denser, and the scaling function approximation in those points becomes more precise. Figure 5.4 represents approximations of the Db2 scaling function (left) and the matching wavelet (right) obtained after two iterations ($j = 2$, step function) and after ten iterations ($j = 10$, the graphs look like continuous functions).

A wavelet graph is generated by the initial set of coefficients $a_{0,n} = 0$ and $b_{0,n} = \delta(n)$ in algorithm (3),

$$a_{-(j+1),n} = \sum_k \big(c(n-2k)\, a_{-j,k} + d(n-2k)\, b_{-j,k}\big), \qquad j = 0, 1, \ldots .$$

The problem of boundary. The DWT algorithm is based on a simple scheme – acting with convolution operators and compression by two, so it is considered standard to use it for signals with a length which is a power of the number two. A signal that does not meet this requirement needs to be prolonged so that its length is a power of the number two prior to the application of the DWT algorithm. It is usually performed in one of the following three ways: zero padding, periodic prolongation or symmetrization, i.e. symmetrical mapping of the border values as a mirror image. Failing of the zero padding is the appearance of an artificial singularity at the border. Failing of the symmetrization is a discontinuity of the first derivative at the border, but this method is generally good for image processing. The positive side of the periodic prolongation is that there are no additional coefficients, but few signals are periodic.

Discrete Wavelet Packet Analysis (DWPA). The variation which can sometimes significantly increase the efficiency of the DWT algorithm by using the its general form is the wavelet packet analysis. As opposed to the DWT algorithm, which decomposes only low-frequency components (approximations), the DWPA algorithm also decomposes the high-frequency components (details). In other words, at each step the low-frequency and high-frequency layers of the signal are decomposed at each step.

5.2 Daubechies wavelets

In 1988 Ingrid Daubechies [19] constructed the entire class of orthonormal wavelet bases Dbr, $r = 1, 2, \ldots$, of compactly supported functions using the analogy with the max-flat filters of signal theory. The scaling function is obtained as a solution of the dilatation equation (3.11) with coefficients $c(k)$ equal to the max-flat filter elements, while the basic wavelet is defined by equation (3.12) with coefficients obtained from formula (3.27). The frequency response (4.5) of the max-flat filter defined by $2r$ dilatation equation coefficients $h(k) = c(k)/\sqrt{2}$ is in the form of

$$(10) \qquad |\hat{h}(\omega)|^2 = \left| \sum_{k=0}^{2r-1} h(k)\, e^{-\imath k\omega} \right|^2 = \left(\frac{1+\cos\omega}{2} \right)^r \hat{q}(\omega).$$

$\hat{q}(\omega)$ is a polynomial of the order r which is chosen to fit orthogonality conditions given by the first relation in (3.26). The first multiplier satisfies the CONDITION A_r (4.22) as it has a zero of the order r in the point $\omega = \pi$. As r increases, the regularity and smoothness properties increase and an approximation error decreases (see §4.4). The choice of the order r defines various Daubechies wavelets that represent a new family of special functions (Figure 5.5). More about relations between Daubechies filters and wavelets is given in §6.4.

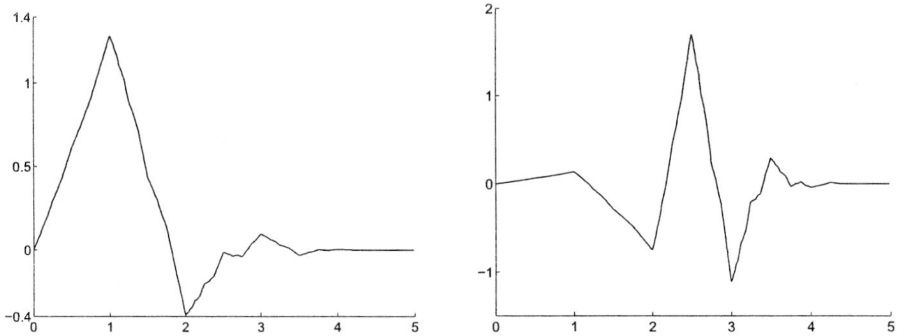

Figure 5.5: Db3 ($r = 3$) scaling function and wavelet

Except the Haar wavelet Db1 (formula 3.35) Daubechies wavelets have no explicit expressions and can only be calculated through recursion. Thus wavelet properties are expressed through the properties of the coefficients $h(k)$, discussed in chapter 4. The length of the compact support for the Dbr wavelet $\psi(x)$ and the scaling function $\varphi(x)$ is $(2r-1)$, while the number of vanishing moments of the wavelet $\psi(x)$ is r. Their regularity increases with the order r, the functions $\psi(x)$ and $\varphi(x)$ belong to the class $C^{\mu r}$, where $\mu \approx 0.2$ for a large r. This means, for example, that if the wavelet $\psi(x)$ should have ten continuous derivatives, the length of its support needs to be around one hundred. Other than the Haar wavelet Db1 the others are not symmetrical, and for some of them the asymmetry is very strong.

Symlet wavelets (Sym$r$) represent a modification of the Daubechies wavelets, done to improve their symmetry. Still, to retain the simplicity of the Daubechies wavelets, they are only almost symmetrical. They are also constructed by the coefficients of the frequency response defined in (10), as there are various ways to group the function (10) to the factors $\hat{h}(\omega)$ and $\overline{\hat{h}(\omega)}$. By the choice of the frequency response $\hat{h}(\omega)$ so that all of its roots by module are smaller than or equal to one, we arrive at the Daubechies wavelet Dbr. By a different choice we arrive to the more symmetry wavelet Sym$r$. Thus the other properties of these wavelets are similar to the Dbr wavelet properties. As mentioned before, a full symmetry cannot be achieved within the frame of the orthonormal wavelet basis with a finite support, other than the Haar Db1 wavelet.

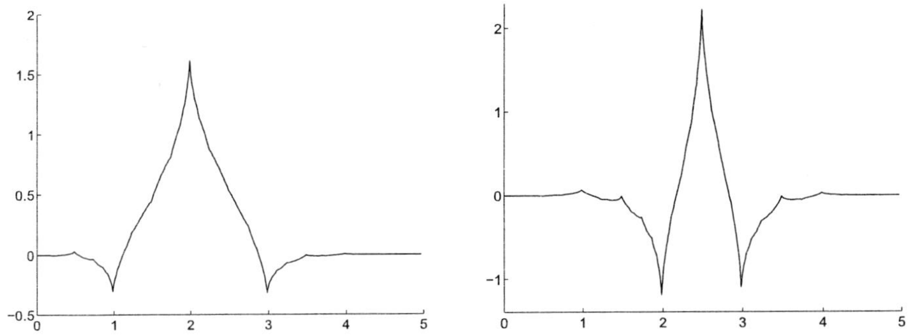

Figure 5.6: The Coiflet scaling function and wavelet

Coiflet wavelets, Coif$r$. They were constructed by Ingrid Daubechies as requested by her colleague Ronald Coifman, whom they were named after. In order to calculate the initial coefficient sequence (5) of the pyramidal algorithm as simply as possible, it is useful to have moments of the scaling function of as high an order as possible equal to zero. By substituting the function $f(x)$ with its Taylor polynomial under this assumption we obtain

$$a_{j,n} = \int f(x)\, 2^{-j/2}\, \varphi(2^{-j}x - n)\, dx = 2^{-j/2} \int f\big(2^j(t+n)\big)\, \varphi(t) 2^j\, dt$$

$$\approx 2^{j/2} \sum_k \frac{f^{(k)}(2^j n)}{k!} \int t^k\, \varphi(t)\, dt = 2^{j/2}\, f(2^j n),$$

because the zero moment of the scaling function $\int \varphi(x)\, dx = 1$. The error is smaller if $\int x^k \varphi(x)\, dx = 0$ for a greater k. The increase of the number of conditions increases the length of the wavelet support. The wavelet Coif$r$ has $2r$, while the corresponding scaling function has $2r-1$ moments equal to zero. Both functions have a support $6r-1$ long. They are less asymmetrical than the wavelets Dbr. Relative to the support length, Coif$r$ is comparable to the wavelets Db$3r$ and Sym$3r$, while relative to the number of vanishing moments it is comparable to the wavelets Db$2r$ and Sym$2r$. They are used in numerical analysis.

More on the wavelets Dbr, Sym$r$ and Coif$r$, as well as the coefficient values defining them for a different r, can be found in [20].

5.3 Biorthogonal wavelets

The orthogonality of a wavelet basis is a very restrictive request, it cannot be fulfilled for the odd number of dilatation equation coefficients (formula 3.27), e.g. the linear spline and matched wavelet (Example 3.3). Besides orthogonal wavelet bases have very strong asymmetry that is an undesirable property in many applications. If we give up of the requirement that the same orthogonal basis is used for the decomposition (analysis) and the reconstruction (synthesis), symmetry is possible. Thus we arrive at symmetric biorthogonal wavelet bases. Instead of one, two sequences of multiresolution spaces are constructed

$$\mathcal{V}_j + \mathcal{W}_j = \mathcal{V}_{j-1}, \qquad \tilde{\mathcal{V}}_j + \tilde{\mathcal{W}}_j = \tilde{\mathcal{V}}_{j-1}.$$

The sums of the spaces are direct (the spaces have no common elements), but in a general case are not orthogonal. Orthogonality exists between spaces of different multiresolutions

$$\mathcal{V}_j \perp \tilde{\mathcal{W}}_j, \qquad \mathcal{W}_j \perp \tilde{\mathcal{V}}_j. \tag{11}$$

If we denote with

$$\tilde{\varphi}_{j,k}(x) = 2^{-j/2}\,\tilde{\varphi}(2^{-j}x - k), \qquad \tilde{\psi}_{j,k}(x) = 2^{-j/2}\,\tilde{\psi}(2^{-j}x - k),$$

scaling functions of dual approximation spaces $\tilde{\mathcal{V}}_j$ and wavelets of dual wavelet spaces $\tilde{\mathcal{W}}_j$, the dilatation (3.11) and wavelet equations (3.12) of two multiresolutions are

$$\begin{aligned} \varphi(x) &= \sum_k h_0(k)\,\varphi(2x-k), &\qquad \tilde{\varphi}(x) &= 2\sum_k f_0(k)\,\tilde{\varphi}(2x-k), \\ \psi(x) &= \sum_k h_1(k)\,\varphi(2x-k), &\qquad \tilde{\psi}(x) &= 2\sum_k f_1(k)\,\tilde{\varphi}(2x-k). \end{aligned} \tag{12}$$

The bases of the spaces that are related by formulae (11) are biorthogonal (formula (2.9)

$$\begin{aligned} (\varphi_{j,k},\,\tilde{\psi}_{j,K}) &= 0, &\qquad (\tilde{\varphi}_{j,k},\,\psi_{j,K}) &= 0, \\ (\psi_{j,k}\,,\,\tilde{\psi}_{J,K}) &= \delta(j-J)\,\delta(k-K), &\qquad (\varphi_{j,k}\,,\,\tilde{\varphi}_{j,K}) &= \delta(k-K). \end{aligned}$$

which leads to the following relations between the coefficients of the equations (12)

$$\begin{aligned} h_1(n) &= (-1)^{n+1}\,f_0(1-n), \\ f_1(n) &= (-1)^{n+1}\,h_0(1-n), \end{aligned} \qquad n = 0, \pm1, \ldots. \tag{13}$$

It is explained how in §6.5.

A consequence of the biorthogonality of these two function systems is that the arbitrary function $g(x) \in \mathcal{L}_2(R)$ can be represented by the decompositions

$$\begin{aligned} g(x) &= \sum_k a_{j,k}\,\tilde{\varphi}_{j,k}(x), &\qquad a_{j,k} &= (g,\,\varphi_{j,k}) = \int g(x)\,\varphi_{j,k}(x)\,dx, \\ g(x) &= \sum_j \sum_k b_{j,k}\,\tilde{\psi}_{j,k}(x), &\qquad b_{j,k} &= (g,\,\psi_{j,k}) = \int g(x)\,\psi_{j,k}(x)\,dx. \end{aligned} \tag{14}$$

The pyramid algorithm (3.32) is defined by the analysis coefficients $h_j(k)$, $j = 0, 1$,

$$a_{j,k} = \sum_l h_0(l - 2k)\, a_{j-1,l}, \qquad b_{j,k} = \sum_l h_1(l - 2k)\, a_{j-1,l},$$

while the inverse pyramid algorithm (3.33) is defined by the synthesis coefficients $f_j(k)$, $j = 0, 1$,

$$a_{j-1,l} = \sum_k \big(f_0(l - 2k)\, a_{j,k} + f_1(l - 2k)\, b_{j,k}\big).$$

The biorthogonal bases can commute roles, so that the tilde functions are used in analysis while the dual ones are used in synthesis. The choice depends on the regularity and number of vanishing moments for both. If $\psi_{j,k}$ and $\tilde{\psi}_{j,k}$ form dual Riesz bases of the wavelets with compact support, then a link exists between the number of vanishing moments of one wavelet and the regularity of the dual wavelet: if $\tilde{\psi} \in C^m$, then $\int x^k \psi(x)\, dt = 0$, $k = 0, \ldots, m$, [20]. At that choice the synthesis by the more regular wavelets is better, in this case $\tilde{\psi}(x)$ as in formulae (14). Moreover, besides the approximation smoothness a higher compression is provided. Due to $(m + 1)$ vanishing moments of the wavelet $\psi(x)$ determining the decomposition coefficients $b_{j,k}$ in (14) most of those coefficients for smooth functions $g(x)$ will be negligible.

EXAMPLE 3. As an illustration, Figure 5.7 shows the dual scaling function (c) and the dual wavelets (d) attached to the roof function (a) and its matching wavelet (b). The coefficients generating these functions are equal to

k	-2	-1	0	1	2
$2\,h_0(k)$	0	1	2	1	0
$4\,f_0(k)$	-1	2	6	2	-1
$4\,h_1(k)$	-1	-2	6	-2	-1
$2\,f_1(k)$	0	-1	2	-1	0

∎

The advantage of biorthogonal over orthogonal bases is in the possibility of constructing symmetrical scaling functions and wavelets. This can be achieved by constructing interpolation scaling functions or by choosing cardinal B-splines as the scaling functions, more on which will be said in the next two sections. All functions, including dual ones, have compact supports and a linear phase. The coefficients are dyadic rational numbers. This is the reason why in equations (12) norming was not done with a coefficient of $\sqrt{2}$, but why the coefficient 2 appears only in dual equations. The failure of these wavelets is that the dual functions have a small smoothness.

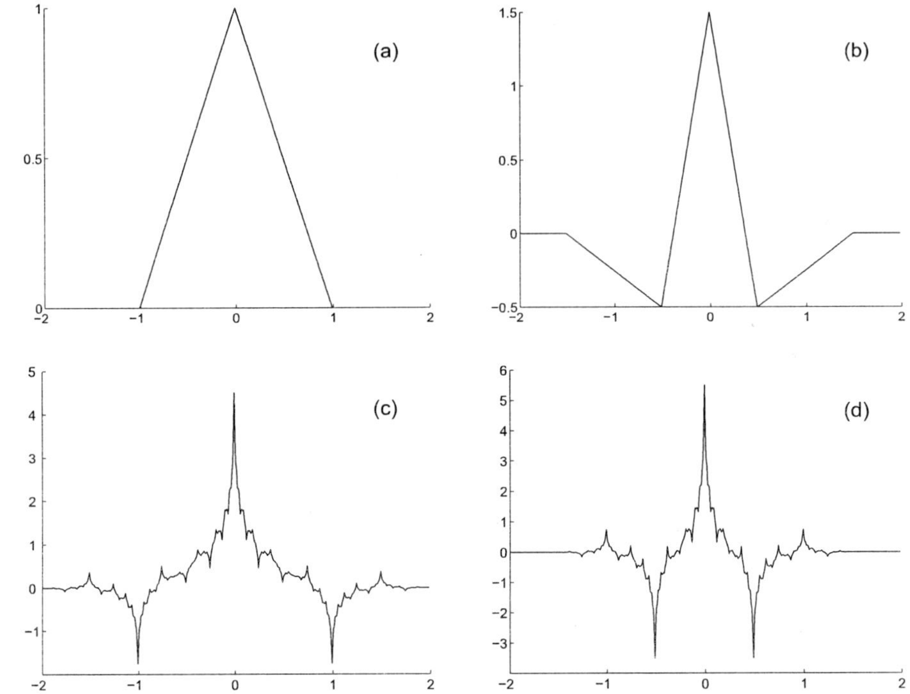

Figure 5.7: Biorthogonal scaling functions (a, c) and wavelets (b, d)

5.4 Cardinal B-splines

Cardinal B-splines represent a family of scaling functions that generate biorthogonal wavelet bases. The cardinal B-spline of the order N is a piecewise N-th degree polynomial with continuous derivatives up to the order $(N-1)$, defined on the integer division and with the interval $[0,\, N+1]$ as a compact support. The cardinal B-spline of the order zero is well known box function $\varphi_0(x) = \aleph_{[0,1)}(x)$, where $\aleph_{[0,1)}(x)$ marks the characteristic function of the interval $[0,1)$. The cardinal B-spline $\varphi_N(x)$ of the order $N \geq 1$ is defined recursively by convolution (2.19)

$$\varphi_N(x) = (\varphi_{N-1} * \varphi_0)(x). \tag{15}$$

Convolution in the physical domain is equivalent to the multiplication in the frequency domain (formula (2.20)), thus we get

$$\hat{\varphi}_N(\omega) = \hat{\varphi}_{N-1}(\omega)\,\hat{\varphi}_0(\omega) = \cdots = \big(\hat{\varphi}_0(\omega)\big)^{N+1}.$$

Since the Fourier transform of the box function equals

$$\hat{\varphi}_0(\omega) = \int_0^1 e^{-\imath\omega t}\, dt = \frac{1 - e^{-\imath\omega}}{\imath\omega},$$

the Fourier transform of the B-spline of the order $N-1$, as convolution of N box functions, is as follows

$$\hat{\varphi}_{N-1}(\omega) = \big(\hat{\varphi}_0(\omega)\big)^N = \left(\frac{1}{\imath\omega}\right)^N (1 - e^{-\imath\omega})^N.$$

According to Example 4.3 the frequency response

$$H_0(z) = \frac{1}{2} + \frac{1}{2} z^{-1}$$

is attached to the box function φ_0 as a scaling function. The frequency response $H_1(z) = H_0^2(z)$ is attached to the roof function (linear spline) $\varphi_1(x) = (\varphi_0 * \varphi_0)(x)$. This is a consequence of the more general statement.

THEOREM 3. *If frequency responses, defined in (4.5),*

$$F(z) = \sum_k f(k)\, z^{-k}, \qquad G(z) = \sum_k g(k)\, z^{-k}, \qquad (z = e^{\imath\omega})$$

correspond to scaling functions $\varphi_f(x)$ *and* $\varphi_g(x)$ *in turn, then the frequency response* $F(z)\, G(z)$ *with the coefficients* $(\mathbf{f} * \mathbf{g})(n)$,

$$F(z)\, G(z) = \sum_k (\mathbf{f} * \mathbf{g})(k)\, z^{-k},$$

corresponds to the scaling function $(\varphi_f * \varphi_g)(x)$,

$$(\varphi_f * \varphi_g)(x) = \sum_k (\mathbf{f} * \mathbf{g})(k)\, (\varphi_f * \varphi_g)(2x - k).$$

Proof: From (2.18) and (4.6) follows

$$\begin{aligned}
\widehat{(\varphi_f * \varphi_g)}(\omega) &= \hat{\varphi}_f(\omega)\, \hat{\varphi}_g(\omega) = F\big(\tfrac{\omega}{2}\big)\, \hat{\varphi}_f\big(\tfrac{\omega}{2}\big)\, G\big(\tfrac{\omega}{2}\big)\, \hat{\varphi}_g\big(\tfrac{\omega}{2}\big) \\
&= \left(\sum_k \sum_l f(k)\, g(l)\, e^{-\imath(k+l)\omega/2}\right) \hat{\varphi}_f\big(\tfrac{\omega}{2}\big)\, \hat{\varphi}_g\big(\tfrac{\omega}{2}\big) \\
&= \left(\sum_m \sum_k f(k)\, g(m-k)\, e^{-\imath m\omega/2}\right) \hat{\varphi}_f\big(\tfrac{\omega}{2}\big)\, \hat{\varphi}_g\big(\tfrac{\omega}{2}\big).
\end{aligned}$$

Taking into consideration that

$$(\mathbf{f} * \mathbf{g})(m) = \sum_k f(k)\, g(m-k), \qquad \widehat{(\varphi_f * \varphi_g)}\big(\tfrac{\omega}{2}\big) = \hat{\varphi}_f\big(\tfrac{\omega}{2}\big)\, \hat{\varphi}_g\big(\tfrac{\omega}{2}\big),$$

one obtains decomposition (4.6) for the function $(\varphi_f * \varphi_g)(x)$,

$$\widehat{(\varphi_f * \varphi_g)}(\omega) = \Big(\sum_m (\mathbf{f} * \mathbf{g})(m)\, e^{-\imath m\omega/2}\Big)\, \widehat{(\varphi_f * \varphi_g)}\big(\frac{\omega}{2}\big)$$

$$= F\big(e^{\imath\omega/2}\big)\, G\big(e^{\imath\omega/2}\big)\, \widehat{(\varphi_f * \varphi_g)}\big(\frac{\omega}{2}\big).$$

∎

By recursion follows that the frequency response $H_{N-1}(z) = H_0^N(z)$ is attached to the B-spline of the order $N-1$ as a scaling function

(16) $$\varphi_{N-1}(x) = \underbrace{(\varphi_0 * \ldots * \varphi_0)}_{\text{N times}}(x).$$

The coefficients of the dilatation equation (4.3) that has the spline of order $N-1$ as a solution, can thus be determined in two ways:

1. as the coefficients of the polynomial $H_{N-1}(z) = \big(\frac{1+z^{-1}}{2}\big)^N$,
2. by the convolution of the vectors $\underbrace{\big(\frac{1}{2}, \frac{1}{2}\big) * \big(\frac{1}{2}, \frac{1}{2}\big) * \cdots * \big(\frac{1}{2}, \frac{1}{2}\big)}_{\text{N times}}$.

EXAMPLE 4. The coefficients of the dilatation equation that determine the linear spline (roof function) $\varphi_1(x)$ are coefficients of the following polynomial per z^{-1},

$$H_1(z) = \Big(\frac{1+z^{-1}}{2}\Big)^2 = \frac{1}{4} + \frac{1}{2}\,z^{-1} + \frac{1}{4}\,z^{-2},$$

but also can be obtained by convolution

$$\begin{array}{ccccc} \big(\frac{1}{2}, \frac{1}{2}\big) & * & \big(\frac{1}{2}, \frac{1}{2}\big) & = & \big(\frac{1}{4}, \frac{1}{2}, \frac{1}{4}\big) \\ \text{box function} & * & \text{box function} & = & \text{roof function} \end{array}$$

∎

Coefficients of the frequency response

(17) $$H_{N-1}(z) = H_0^N(z) = \big(\frac{1+z^{-1}}{2}\big)^N = \sum_{k=0}^{N} \frac{1}{2^N} \binom{N}{k} z^{-k},$$

define the dilatation equation (4.3) with the spline of the order $N-1$ as a solution,

$$\varphi_{N-1}(x) = 2^{1-N} \sum_{k=0}^{N} \binom{N}{k}\, \varphi_{N-1}(2x-k).$$

The differentiation of the recurrent formula (15), defining a spline with an order greater by one than the previous one,

$$\varphi_{N-1}(x) = (\varphi_{N-2} * \varphi_0)(x) = \int_{-\infty}^{\infty} \varphi_0(t)\varphi_{N-2}(x-t)\,dt = \int_0^1 \varphi_{N-2}(x-t)\,dt,$$

yields a recurrent formula for calculating the spline derivative

$$\varphi'_{N-1}(x) = \int_0^1 \varphi'_{N-2}(x-t)\,dt = \varphi_{N-2}(x) - \varphi_{N-2}(x-1). \tag{18}$$

By further differentiation $N-1$ times of the recurrent relation (18), we get the expressions for higher order derivatives of the B-spline

$$\varphi''_{N-1}(x) = \varphi'_{N-2}(x) - \varphi'_{N-2}(x-1) = \varphi_{N-3}(x) - 2\,\varphi_{N-3}(x-1) + \varphi_{N-3}(x-2)$$

$$\vdots$$

$$\varphi^{(N-1)}_{N-1}(x) = \sum_{k=0}^{N} (-1)^k \binom{N}{k} \varphi_0(x-k).$$

The above conclusion will be summed up by the following theorem.

THEOREM 4. *The cardinal B-spline $\varphi_{N-1}(x)$ of the order $N-1$, determined by the convolution of N box functions, is a piecewise polynomial of the degree $N-1$. The finite discontinuities of the $(N-1)$-th derivative at the points $x = 0, 1, \ldots, N$, are equal to the binomial coefficients with an alternating sign change,*

$$\lim_{x\to k+} \varphi^{(N-1)}_{N-1}(x) - \lim_{x\to k-} \varphi^{(N-1)}_{N-1}(x) = (-1)^k \binom{N}{k}, \qquad k = 0, \ldots, N.$$

∎

EXAMPLE 5. The cubic B-spline $\varphi_3(x)$ is the solution of the dilatation equation (4.3) with the coefficients (Example 3.5)

$$\mathbf{h} = \left(\frac{1}{16}, \ \frac{4}{16}, \ \frac{6}{16}, \ \frac{4}{16}, \ \frac{1}{16}\right).$$

Five coefficients determine that the length of its compact support is four unit intervals (Theorem 4.1). The point $z = -1$ is the zero of the order four of its frequency response

$$H_3(z) = \left(\frac{1+z^{-1}}{2}\right)^4 = \frac{1}{16} + \frac{4}{16}z^{-1} + \frac{6}{16}z^{-2} + \frac{4}{16}z^{-3} + \frac{1}{16}z^{-4}.$$

It means the cubic B-spline has the greatest possible smoothness and has the greatest possible order of the accuracy $r = 4$ for filters five long (see §4.4).

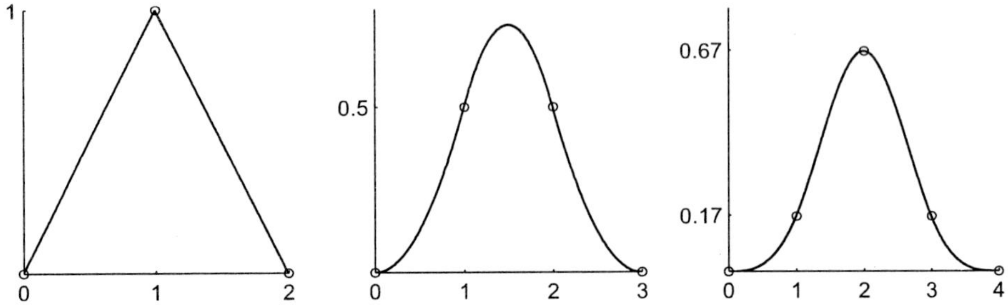

Figure 5.8: Linear, square and cubic spline

This spline is a convolution of the four box functions,

$$\varphi_3(x) = (\varphi_0 * \varphi_0 * \varphi_0 * \varphi_0)(x).$$

The result of the first convolution is the linear spline $\varphi_1(x)$ (roof function), of the second convolution the square spline $\varphi_2(x)$ and of the final one the cubic spline $\varphi_3(x)$ (Figure 5.8). They belong to the classes of continuous functions C^0, C^1 and C^2, in turn. The fourth derivative of the cubic spline in a generalized sense is the sum of delta functions. ∎

Let us now analyze the spline bases through frequency responses (17) attached to splines. The inner product of the spline with its translation equals a spline of a higher order,

$$a(n) = \int_{-\infty}^{\infty} \varphi_{N-1}(x)\, \varphi_{N-1}(x+n)\, dx = \varphi_{2N-1}(N+n),$$

because the integral is convolution of N box functions with N new box functions shifted by n, whereas $2N$ box functions generate a spline of the order $2N-1$. The vector $\mathbf{a}$, the elements of which are inner products $a(n)$, is the solution of the equation $\mathbf{a} = T\mathbf{a}$ (§4.5). The operator T (formula (4.14)) is determined by the product

$$H_{N-1}(z)\, H_{N-1}(z^{-1}) = \left(\frac{1+z^{-1}}{2}\right)^N \left(\frac{1+z}{2}\right)^N = z^N \left(\frac{1+z^{-1}}{2}\right)^{2N} = z^N\, H_{2N-1}(z),$$

Thus, the matrix T for a spline of the order $N-1$ is identical to the matrix M (formula (4.13)) for the spline of the order $2N-1$. Elements of the eigenvector of the matrix M are values of the $2N-1$ order spline at integer points. The same eigenvector, as the eigenvector of the matrix T, has as its elements inner products of the spline of the order $N-1$. Since for all splines $\varphi(x) \geq 0$, all inner products $a(n) \geq 0$. The sum of inner products equals one, because the sum of the scaling function values at integer points equals one (formula (4.21)).

The frequency response $H_{N-1}(z)$ has a zero of the order N for $z = -1$, i.e. it has N zeroes in the point $\omega = \pi$. According to that which was proven in §4.4 the polynomials $1, x, x^2, \dots, x^{N-1}$ can be reproduced by the splines of the order $N-1$, thus the approximation accuracy using them is of the order N. A wavelet, generated by the scaling function $\varphi_{N-1}(x)$, will have N vanishing moments. The question is which wavelets correspond to splines?

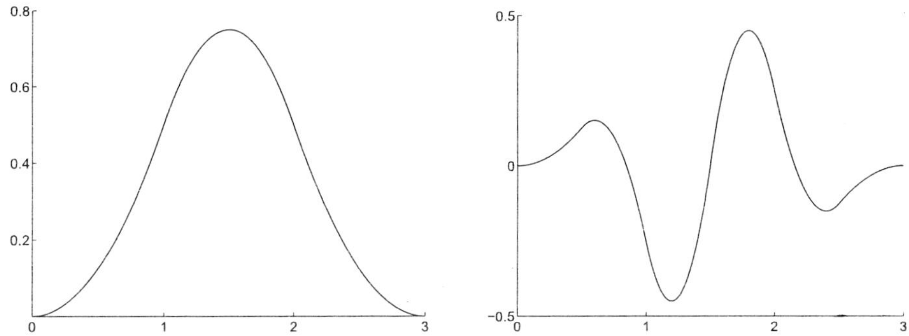

Figure 5.9: Square spline and the attached wavelet

The basis $\{\varphi_N(x-k)\}$ is not orthogonal (except for $N = 0$, box function). The wavelet attached to the square spline, determined by the coefficients $d(k)$ which satisfy relation (3.27), is represented on Figure 5.9. It is not possible to attach wavelets in this way to the linear and the cubic B-spline because they are defined by coefficients vectors with odd lengths ($N = 3$ for the linear and $N = 5$ for the cubic spline). We are looking for another way to construct the wavelets $\psi_N(x-k)$ orthogonal to the spline $\varphi_N(x)$. The spaces they generate has to fulfil conditions

$$\mathcal{V}_0 \perp \mathcal{W}_0, \qquad \mathcal{V}_0 \oplus \mathcal{W}_0 = \mathcal{V}_{-1}.$$

Thus we arrive at spaces that are orthogonal, the bases of which are not orthogonal. *Semiorthogonal wavelets* are basis functions orthogonal only if they belong to different scales: $\psi_N(2^j x - k)$ and $\psi_N(2^J x - l)$ are orthogonal if $j \neq J$. This is a direct consequence of the presumed orthogonality of the spaces, because $\mathcal{W}_j$ is orthogonal to $\mathcal{V}_j$, thus to all subspaces $\mathcal{W}_{j+1}, \mathcal{W}_{j+2}, \dots$, as well, contained within it. On the same scale, semiorthogonal wavelets are not orthogonal in the general case.

EXAMPLE 6. The semiorthogonal wavelet corresponding to the roof function is

$$\psi_1(x) = \frac{1}{12}\big(\varphi(2x) - 6\,\varphi(2x-1) + 10\,\varphi(2x-2) - 6\,\varphi(2x-3) + \varphi(2x-4)\big).$$

This wavelet is orthogonal to all linear splines $\varphi_1(x-n)$. It is not orthogonal to all own translations $\psi_1(x-n)$, but it is orthogonal to all the wavelets $\psi_1(2^j x - n)$, $j \neq 0$ (Figure 5.10). ∎

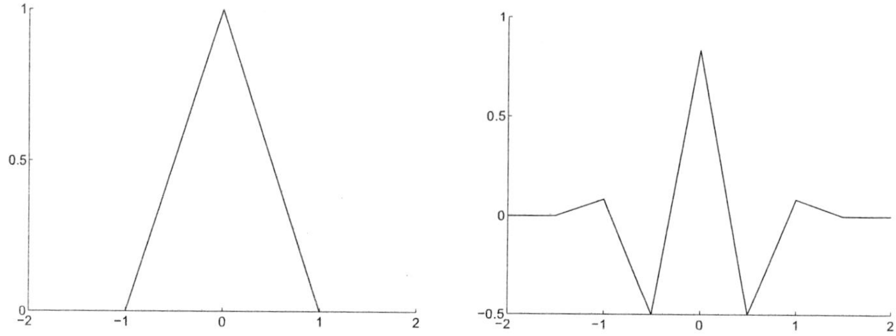

Figure 5.10: The linear spline and attached semiorthogonal wavelet

Orthogonal scaling function and wavelet bases (mutually orthogonal) cannot be constructed for splines. The orthogonality problem can be overcome by the construction of biorthogonal bases. The frequency response of the B-spline of the order $N-1$ (formula (17) for $z = e^{\imath\omega}$), with no delay equals

$$\hat{h}_{N-1}(\omega) = \left(\cos\frac{\omega}{2}\right)^N, \quad \text{for} \quad N = 2l,$$

or

$$\hat{h}_{N-1}(\omega) = e^{-\imath\omega/2}\left(\cos\frac{\omega}{2}\right)^N, \quad \text{for} \quad N = 2l+1,$$

thus the B-spline $\varphi_{N-1}(x)$ is centered around zero or $1/2$ (depends on N). The frequency response of the dual basis, defining the coefficients $f_0(k)$ in formula (12), is

$$\hat{f}_{N-1}(\omega) = \left(\cos\frac{\omega}{2}\right)^{\tilde{N}} \sum_{m=0}^{l+\tilde{l}-1} \binom{l+\tilde{l}-1-m}{m} \left(\sin^2\frac{\omega}{2}\right)^m, \quad \text{for} \quad \tilde{N} = 2\tilde{l}$$

or

$$\hat{f}_{N-1}(\omega) = e^{-\imath\omega/2}\left(\cos\frac{\omega}{2}\right)^{\tilde{N}} \sum_{m=0}^{l+\tilde{l}} \binom{l+\tilde{l}-m}{m} \left(\sin^2\frac{\omega}{2}\right)^m, \quad \text{for} \quad \tilde{N} = 2\tilde{l}+1.$$

The algorithm described in §5.2 is then applied.

Finally, an orthogonal basis using the B-spline bases can be constructed by the Gram-Schmidt orthogonalization process.

Battle–Lemarie wavelets. [13] They are constructed by the orthogonalization of the B-spline basis. Even though B-splines have compact supports, the orthogonalization of the basis yields scaling functions with infinite supports. Using the Fourier transform of the scaling function, the wavelet and its attached coefficients are determined. The wavelet, just like the scaling function, has an infinite support, but it decreases towards zero exponentially. The wavelet attached to the spline of the order N has derivatives up to the order $N-1$.

5.5 Interpolation wavelets

Interpolation wavelets represent an extreme case of biorthogonal wavelets, when the dual scaling function, and thus the dual wavelet, are represented by Dirac functions. They have many nice properties such as

- symmetry;
- the coefficients in the scaling function representation are values of a physical quantity in the dyadic points;
- the moments and coefficients in the representation are easy to determine;
- the basis functions can be determined without solving an eigenvalue problem;
- they provide a more accurate approximation.

Due to these properties they are especially convenient for modeling with partial differential equations.

Conceptually, these are the simplest wavelets possible. Their construction is based on the interpolation recursion, used to construct a continuous function given by its values in a finite number of points. Let the function $y(x)$ be given by its values $y_l = y(x_l)$ in regularly distributed points x_l, where we shall assume, without reducing the generality, that x_l are integers. We shall call this grid

level 0 $\qquad (x_l, y_l), \qquad x_l = l \in \mathcal{Z}.$

Let us calculate the approximate values of the function in the centers of integer intervals using interpolation polynomials of the degree $(M-1)$. The degree of the interpolation polynomials should be odd, meaning that M is even, so that an equal number $M/2$ of interpolation nodes should be distributed on both sides of the point where we calculate the interpolated value.

EXAMPLE 7. If we use a third order polynomial $(M = 4)$, the approximate value of the function $y_{i+3/2}$ at the center point of the interval determined by the four subsequent interpolation nodes x_{i+k}, $k = 0, 1, 2, 3$, equals [37]

$$
\begin{aligned}
y(x_{i+3/2}) \approx y_{i+3/2} &= \sum_{k=0}^{3} \left(\prod_{\substack{l=0 \\ l\neq k}}^{3} \frac{x_{i+3/2} - x_{i+l}}{x_{i+k} - x_{i+l}} \right) y_{i+k} \\
&= -\frac{1}{16} y_i + \frac{9}{16} y_{i+1} + \frac{9}{16} y_{i+2} - \frac{1}{16} y_{i+3}.
\end{aligned} \tag{19}
$$

∎

By calculating the approximate values of the function using the same algorithm in all centers of integer intervals, we arrive at the approximate values of the function $y(x)$ at the following level,

$$level\,1 \qquad (x_l,\, y_l), \qquad x_l = 2^{-1}\,l \in \mathcal{Z},$$

containing twice as many points as *level 0*. In the next step we start from this new set of data, and using the algorithm described calculate the approximate function values at the centers of the new (the quarters of the initial) intervals, defining the grid

$$level\,2 \qquad (x_l,\, y_l), \qquad x_l = 2^{-2}\,l \in \mathcal{Z}.$$

It is obvious that each subsequent level contains twice as much information about the function relative to the previous one, representing multiresolution.

By infinite repetition of the interpolation a quasi-continuous function, defined in all dyadic points $x_l = 2^{-j}\,l,\ j \to \infty$, is generated. This function has a continuous extension $\breve{y}(x),\ x \in \mathcal{R}$, defining the mapping of the initial set to the continuous function ([22]),

$$(20) \qquad \breve{y}(x) = I_M(\mathbf{y}), \qquad \mathbf{y} = \{y_l\}, \quad l \in \mathcal{Z}.$$

The index M is higher by one than the degree of the interpolation polynomial used in the recursion.

The interpolation scaling function $\varphi(x)$ is a continuous expansion of the limiting function determined by the interpolation recursion, when the initial values are given by the unit vector $\mathbf{e_0}$, $e_{0,l} = \delta(l)$. For a chosen M, in accordance with (20),

$$\varphi(x) = I_M(\mathbf{e_0}).$$

Based on the construction it is clear that the compact support of the function $\varphi(x)$ is the interval $[-(M-1),\,(M-1)]$. The translation of the scaling function $\varphi(x-k)$ is determined by the initial data given by the k-th unit vector on a grid with a step of one,

$$\varphi(x-k) = I_M(\mathbf{e_k}), \qquad e_{k,l} = \delta(k-l), \quad k, l \in \mathcal{Z}.$$

An arbitrary set of data $\mathbf{y} = \{y_k\}$ can be represented by a linear combination of unit vectors $\mathbf{y} = \sum_k y_k\,\mathbf{e_k}$. Since interpolation is a linear process, it follows that the limiting function $\breve{y}(x)$, defined by the data $\mathbf{y}$, can be expressed using an interpolation scaling function and its translations,

$$\breve{y}(x) = I_M(\mathbf{y}) = I_M\big(\sum_k y_k\,\mathbf{e_k}\big) = \sum_k y_k\,I_M(\mathbf{e_k}) = \sum_k y_k\,\varphi(x-k).$$

If the data represents the values of function $y(x)$ which is a polynomial with a degree not greater than the degree of the interpolation polynomial used in the algorithm, this function will be accurately reconstructed. Every smooth function can be approximated well using a polynomial, thus the interpolation scaling function $\varphi(x)$ represents a good choice for generating a wavelet family yielding a high accurate approximation.

In recursion (20), we have started from *level 0*, where the data was given in integer points. If we start from level j, where the data is given in dyadic points $x_l = 2^{-j}\,l$ by the unit vector at that level $\mathbf{e_k}^{(j)}$, $e^{(j)}_{k,l} = \delta(k-l)$ on a grid with a step of 2^{-j}, i.e. if we start from the data

$$\left(2^{-j}\,l,\, e^{(j)}_{k,l}\right), \qquad l \in \mathcal{Z},$$

by the mapping (20) we shall again arrive to the scaling function, compressed by a factor of 2^j,

$$\varphi(2^j x - k) = I_M\left(\mathbf{e_k}^{(j)}\right). \tag{21}$$

We want to establish a relation between the limiting functions given by the data on two subsequent levels, i.e. to arrive at the dilatation equation for the interpolation scaling function.

EXAMPLE 8. In order to arrive at the relation being sought more easily then in a general case, let us observe what it is like for an interpolation process defined by a third degree polynomial, as described in Example 7. Starting from the data $\mathbf{e_0}$, the interpolation formula (19) yields the values of the function $\varphi(x)$ at the centers of the unit intervals. Values different from zero are gained for $i=-1$ and $i=0$,

$$\begin{aligned}\varphi\left(\frac{1}{2}\right) &= -\frac{1}{16}\,e_{0,-1} + \frac{9}{16}\,e_{0,0} + \frac{9}{16}\,e_{0,1} - \frac{1}{16}\,e_{0,2}\\ &= -\frac{1}{16}\cdot 0 + \frac{9}{16}\cdot 1 + \frac{9}{16}\cdot 0 - \frac{1}{16}\cdot 0 = \frac{9}{16}\end{aligned}$$

$$\begin{aligned}\varphi\left(\frac{3}{2}\right) &= -\frac{1}{16}\,e_{0,0} + \frac{9}{16}\,e_{0,1} + \frac{9}{16}\,e_{0,2} - \frac{1}{16}\,e_{0,3}\\ &= -\frac{1}{16}\cdot 1 + \frac{9}{16}\cdot 0 + \frac{9}{16}\cdot 0 - \frac{1}{16}\cdot 0 = -\frac{1}{16}.\end{aligned}$$

Taking into consideration the symmetry and the values in integer points given at *level* 0, the function $\varphi(x)$ is given at *level* 1 (in the points $x_k = k/2$) by the following table

k	-3	-2	-1	0	1	2	3
$\varphi(k/2)$	$-1/16$	0	$9/16$	1	$9/16$	0	$-1/16$

The other values are equal to zero, confirming the earlier conclusion that the support is compact, whereas in this case it is equal to the six basic intervals. The data in the table can be written in the following form, using unit vectors $\mathbf{e_k}^{(1)}$ attached to the *level* 1, i.e. to the division of the interval into halves

$$\Phi^{(1)} = -\frac{1}{16}\,\mathbf{e_{-3}}^{(1)} + \frac{9}{16}\,\mathbf{e_{-1}}^{(1)} + \mathbf{e_0}^{(1)} + \frac{9}{16}\,\mathbf{e_1}^{(1)} - \frac{1}{16}\,\mathbf{e_3}^{(1)}. \tag{22}$$

This data defines the same recursion as vector $\mathbf{e_0}$, leading to the scaling function

$$I_4(\Phi^{(1)}) = \varphi(x). \tag{23}$$

By substituting (22) into (23), taking into consideration (21), we arrive at the recurrent relation which is sought

$$\varphi(x) = -\frac{1}{16}\,\varphi(2x+3) + \frac{9}{16}\,\varphi(2x+1) + \varphi(2x) + \frac{9}{16}\,\varphi(2x-1) - \frac{1}{16}\,\varphi(2x-3), \tag{24}$$

representing the dilatation equation for the interpolation scaling function determined by the cubic interpolation $(M = 4)$. It is obvious that the coefficients of the equation have values of $\varphi(l/2),\, l \in \mathcal{Z}$, (see table), which is valid for the general case as well. ∎

The interpolation of the order $(M-1)$ (with M interpolation nodes) defines the scaling function which is the solution of the dilatation equation

$$\varphi(x) = \sum_{k=-M+1}^{M-1} \varphi(k/2)\,\varphi(2x-k).$$

The coefficients $\mathbf{h}_0$ of the analysis frequency response are the scaling function values in points of the *level* 1,

$$h_0(k) = \varphi(k/2), \qquad k = 0, \pm 1, \ldots, \pm(M-1).$$

By comparing expressions (19) and (24), we can see that these coefficients are, in fact, coefficients of the interpolation formula. Let us arrive at the expression used to calculate them in the general case for interpolation with M nodes (M is even). The Lagrange interpolation polynomial is invariant in relation to the linear substitution of the variable $x = (t + M - 1)/2$, used to map the nodes $x_k = k$, $k = 0, \ldots, M-1$, into the nodes $t_k = 2k - (M-1)$,

$$L(t) = \sum_{k=0}^{M-1} \left(\prod_{\substack{l=0 \\ l \neq k}}^{M-1} \frac{t - t_l}{t_k - t_l} \right) f_k = \sum_{k=0}^{M-1} c_k(t)\, f_k. \tag{25}$$

The center of the interpolation interval is mapped by this substitution into the point $t = 0$, thus the dilatation equation coefficients are determined by the interpolation formula coefficients (25),

$$c_k(0) = \prod_{\substack{l=0 \\ l \neq k}}^{M-1} \frac{-t_l}{t_k - t_l} = \prod_{\substack{l=0 \\ l \neq k}}^{M-1} \frac{M-1-2l}{2(k-l)} = \frac{1}{2^{M-1}} \prod_{\substack{l=0 \\ l \neq k}}^{M-1} \frac{M-1-2l}{k-l}, \qquad k = 0, \ldots, M-1,$$

by the following relations

$$h_0(0) = 1, \qquad \begin{array}{c} h_0(2n) = 0, \quad h_0(2n-1) = c_{M/2-n}, \\ h(-n) = h(n) \end{array} \qquad n = 1, \ldots, M/2.$$

Now we shall determine the dual scaling function $\tilde{\varphi}(x)$. From the biorthogonality condition of the bases given by the functions $\varphi(x)$ and $\tilde{\varphi}(x)$,

$$\int \tilde{\varphi}(x)\, \varphi(x-n)\, dx = \delta(n),$$

and the property of the interpolation scaling function that $\varphi(n) = \varphi(-n) = \delta(n)$, $n \in \mathcal{Z}$, it follows that the dual scaling function is Dirac function (Example 2.4),

$$\int \tilde{\varphi}(x)\, \varphi(x-n)\, dx = \varphi(-n) \qquad \longrightarrow \qquad \tilde{\varphi}(x) \equiv \delta(x). \tag{26}$$

The dilatation equation of the dual scaling function is therefore the dilatation equation having as its solution the function $\delta(x)$,

$$\tilde{\varphi}(x) = 2\,\tilde{\varphi}(2x), \qquad x \in \mathcal{R}. \tag{27}$$

From (27) we conclude that the coefficients of the synthesis frequency response are

$$f_0(0) = 1, \qquad f_0(n) = 0, \quad n \in \mathcal{Z} \backslash \{0\}.$$

The wavelet coefficients are determined by the relations (13),

$$\begin{array}{c} h_1(1) = f_0(0) = 1, \qquad h_1(n) = 0, \quad n \neq 1, \\ f_1(n) = (-1)^{n+1} h_0(1-n) \end{array} \qquad n \in \mathcal{Z}.$$

Substituting these coefficients into the equation (12), we arrive at the matching wavelet

$$\psi(x) = \varphi(2x-1) = \varphi_{1,1}(x), \qquad \longrightarrow \qquad \psi_{j,k}(x) = \varphi_{j+1,2k+1}(x). \tag{28}$$

The dual wavelet is determined by the coefficients $f_1(n)$.

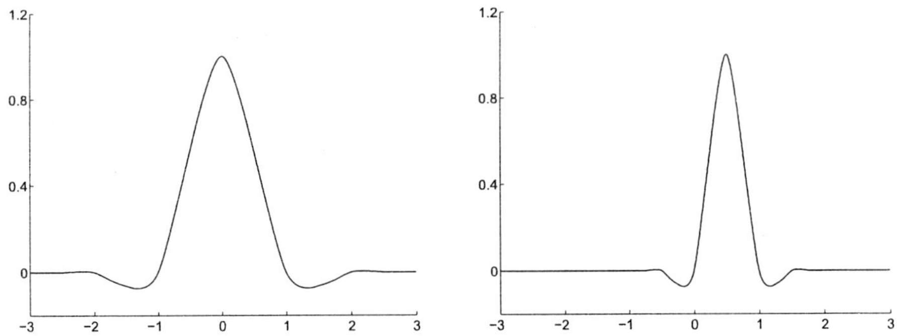

Figure 5.11: The interpolation scaling function and wavelet for $M = 4$

EXAMPLE 9. The coefficients for the interpolation wavelet arrived at in example 8 are given by the following table

k	-4	-3	-2	-1	0	1	2	3	4
$h_0(k)$	0	$-1/16$	0	$9/16$	1	$9/16$	0	$-1/16$	0
$f_0(k)$	0	0	0	0	1	0	0	0	0
$h_1(k)$	0	0	0	0	0	1	0	0	0
$f_1(k)$	0	0	$1/16$	0	$-9/16$	1	$-9/16$	0	$1/16$

The graphs of the scaling function and wavelet are given on Figure 5.11. The equation of the dual wavelet is, according to (12),

$$\tilde{\psi}(x) = 2\left(\frac{1}{16}\,\tilde{\varphi}(2x+2) - \frac{9}{16}\,\tilde{\varphi}(2x) + \tilde{\varphi}(2x-1) - \frac{9}{16}\,\tilde{\varphi}(2x-2) + \frac{1}{16}\,\tilde{\varphi}(2x-4)\right),$$

or, having in mind (26) and (27),

$$\tilde{\psi}(x) = \frac{1}{16}\,\delta(x+1) - \frac{9}{16}\,\delta(x) + \delta(x-1/2) - \frac{9}{16}\,\delta(x-1) + \frac{1}{16}\,\delta(x-2).$$

The coefficients defining the interpolation wavelets for $M = 6$, 8 and 10 can be found in [25].

5.6 Second generation wavelets

The wavelets mentioned so far, generated by translations and dilatations of a single or several basic functions are called first generation wavelets (classic wavelets). Since these operations represent algebraic operations in the frequency domain, the basic tool for their construction is the Fourier transform. There are a number of problems, such as problems defined on intervals, curves, surfaces or manifolds, where the Fourier transform cannot be applied, and thus neither can classic wavelets. Classic wavelets also need to be modified when used for solving problems defined by irregular grids or where the inner product with a weight function needs to be used. Wavelets attached to problems not allowing for translation and dilatation are called *second generation wavelets.* Coefficients that correspond to these wavelets can depend on the resolution level. It is clear that working with non-constant coefficients is more complex.

The basic idea of wavelet transform is to use a correlation existing in most signals in order to construct a good approximation with few addends. A correlation is a typical local property in space (time) and frequency, meaning that neighboring data and frequencies are far more correlated than those further moved from each other. In transformation with classic wavelets the basic tool for the space(time)–frequency localization is the Fourier transform, which cannot be applied to more complex geometries. However, localization can be performed in the physical domain (space, time etc.), which is the essence of the so-called "lifting" algorithm. This algorithm was primarily developed for constructing second generation wavelets, but it is used successfully also for the construction of biorthogonal wavelets [21, 46].

Lifting generalizes the idea of multiresolution to spaces that are not invariant relative to translation and dilatation, thus enabling users to create wavelets according to their needs and to speed up wavelet transform. The basic idea is to use the correlation between neighboring data in the signal. The way in which this is performed shall be illustrated on a simple example of constructing biorthogonal wavelets.

EXAMPLE 10. Let us divide the discrete signal $\mathbf{x} = \{x_k\}_{k\in Z}$ into two disjunctive sets, so that the elements with even indices belong to the first one $\mathbf{x}_a = \{x_{2k}\}$, while the elements with an odd index belong to the second set $\mathbf{x}_b = \{x_{2k+1}\}$. This transformation, usually called the Lazy wavelet transform, does not achieve a reduction in the amount of data used to register a signal because the data in a general case significantly differs from zero and is not negligible (do not alow compression). By presuming the correlation of subsequent data, it is natural to suppose that the signal element with an odd index can approximately be expressed as the arithmetic mean of its neighboring data, i.e. signal elements with even indices. Instead of the signal elements with odd indices, we shall memorize in the signal $\mathbf{b}$ the differences of the approximation from the values themselves,

$$b_k = x_{2k+1} - \frac{1}{2}\left(x_{2k} + x_{2k+2}\right), \qquad (29)$$

that corresponds to the wavelet coefficients. They register high frequencies. If the signal is linear, all coefficients b_k will be zero. In the general case, since these coefficients measure the deviation of the given signal from a linear one, most of them can be expected to be negligible, allowing for good compression.

The data in the signal $\mathbf{x}_a$ describes lower frequencies and corresponds to the approximation. There is half of the number of elements respecting to the input signal. In order to preserve the mean value of the approximation $\mathbf{x}_a$ on each resolution level, we shall correct it at each level by adding the quarter of the sum of the neighboring details to the approximation,

$$a_k = x_{2k} + \frac{1}{4}\left(b_{k-1} + b_k\right).$$

The signal mean value is preserved because, according to (29), the approximation equals

$$\begin{aligned} a_k &= x_{2k} + \frac{1}{4}\left(x_{2k-1} - \frac{1}{2}\left(x_{2k-2} + x_{2k}\right) + x_{2k+1} - \frac{1}{2}\left(x_{2k} + x_{2k+2}\right)\right) \\ &= \frac{3}{4}x_{2k} + \frac{1}{4}\left(x_{2k-1} + x_{2k+1}\right) - \frac{1}{8}\left(x_{2k-2} + x_{2k+2}\right). \end{aligned}$$

Thus the sum of approximations on this level equals half the sum of the approximations on the previous level,

$$\sum_k a_k = \frac{3}{4}\sum_k x_{2k} + \frac{1}{4}\sum_k (x_{2k-1} + x_{2k+1}) - \frac{1}{8}\sum_k (x_{2k-2} + x_{2k+2}) = \frac{1}{2}\sum_k x_k.$$

On this level there are half of the number of elements respecting to the previous level.

The procedure described represents a single step in the lifting algorithm, yielding the approximation $\mathbf{a}$ and the detail $\mathbf{b}$ at the output. The approximation $\mathbf{a}$ represents the input signal $\mathbf{x}$ for the next step. By the subsequent application of the algorithm described on the signal $\mathbf{x}$, i.e. $\mathbf{a}$ determined by the previous iteration, we arrive at more coarse approximations and new details. The result of the application of the lifting algorithm is the approximation of the input signal $\mathbf{x}$ at the last (the most coarse) resolution level $\mathbf{a}^{(J)}$ and details at all levels $\mathbf{b}^{(j)}$, $j = 1, \ldots, J$.

∎

Let us generalize the procedure described in Example 10. A single step of the lifting algorithm consists of three operations: split, predict and update.

- *Split.* The initial signal $\mathbf{x}$ is split into two disjunctive sets of data $\mathbf{x}_a$ and $\mathbf{x}_b$,

$$\mathbf{x}_a \cup \mathbf{x}_b = \mathbf{x}, \qquad \mathbf{x}_a \cap \mathbf{x}_b = \emptyset.$$

- *Predict.* The data from one set is used to predict the data from the other set; for example, the data from the set $\mathbf{x}_b$ is predicted using the prediction operator $P(\mathbf{x}_a)$ (in Example 10, the prediction operator P is the arithmetic mean). This enables the initial set of data $\mathbf{x}$ to be replaced by the subset $\mathbf{x}_a$. In a general case the operator P cannot be constructed so that it determines the set $\mathbf{x}_b$ based on the set $\mathbf{x}_a$ accurately, thus the deviations (details) are defined by the expression

$$\mathbf{b} = \mathbf{x}_b - P(\mathbf{x}_a).$$

The subset $\mathbf{b}$ shows how much the prediction deviates from the accurate data. For a well defined operator P, most of the elements of this set will be numbers close or equal to zero, enabling good data compression.

- *Update.* Some of the properties of the initial set $\mathbf{x}$ (in Example 10 it is the property of maintaining the mean value) are updated to the new, smaller set of data after the correction performed by the details,

$$\mathbf{a} = \mathbf{x}_a + U(\mathbf{b}).$$

U is the updating operator (in Example 10 the operator U is half of the arithmetic mean).

Thus, after a single step of the lifting algorithm we arrive at two signals $\mathbf{a}$ and $\mathbf{b}$, each half the length of the input signal. $\mathbf{a}$ is the input signal for the next step, while the details $\mathbf{b}$ can, with a good choice of the prediction operator P, be efficiently compressed. The procedure described is repeated at each step of the lifting algorithm, providing the approximation $\mathbf{a}^{(j)} = \{a_{j,k}\}_{k\in\mathcal{Z}}$ and detail $\mathbf{b}^{(j)} = \{b_{j,k}\}_{k\in\mathcal{Z}}$ at the j-th step. The algorithm is fully analogue to the pyramidal algorithm (§3.3), thus the same notation is used. The inverse transformation is, obviously, performed very simply with the operations reverse to those used in direct transformation (subtracting the update information, adding the predict information and merge the obtained even and odd samples).

An important property of the lifting algorithm is that the operations can be unified, so that this algorithm can be used for the modification of existing wavelet transforms. If H_0 is the scaling function and H_1 the wavelet frequency responses of the analysis, and F_0 and F_1 their dual pair (in accordance with §5.3), the prediction operator will modify the initial responses in the following way

$$\tilde{H}_0 = H_0, \quad \tilde{H}_1 = H_1 - P\,H_0, \qquad \tilde{F}_0 = F_0 + P\,F_1, \quad \tilde{F}_1 = F_1, \tag{30}$$

while the action of the updating operator on the initial responses is expressed by the relations

$$\tilde{H}_0 = H_0 + U\,H_1, \quad \tilde{H}_1 = H_1, \qquad \tilde{F}_0 = F_0, \quad \tilde{F}_1 = F_1 - U\,F_0. \tag{31}$$

When the initial responses represent Lazy wavelet transform (separation of the elements with even and odd indices), the new biorthogonal filters are, based on (30) and (31),

$$\tilde{H}_0 = I + U\,(I - P), \quad \tilde{H}_1 = I - P, \qquad \tilde{F}_0 = I + P, \quad \tilde{F}_1 = I - U\,(I + P),$$

where I is the identity operator.

EXAMPLE 11. We shall correlate the lifting algorithm from Example 10 with filters and wavelets. Let φ be the roof function, $\varphi(x) = \max\{0,\, 1 - |x|\}$ that is used to defined the piecewise linear approximation at every resolution level j

$$A^{(j)}(x) = \sum_k a_{j,k}\, \varphi_{j,k}(x), \qquad \varphi_{j,k}(x) = \varphi(2^{-j}x - k). \tag{32}$$

The approximation difference at two successive levels will be represented by the wavelets

$$A^{(j-1)}(x) - A^{(j)}(x) = \sum_k b_{j,k}\, \psi_{j,k}(x), \qquad \psi_{j,k}(x) = \psi(2^{-j}x - k). \tag{33}$$

We shall assume that no approximation updating was performed, i.e. that $a_{0,k} = a_{-1,2k}$. To keep the formulae simple, we have left out the norming coefficient $2^{-j/2}$ of the scaling function and wavelet, and we have taken the roof function symmetrical to the origin for the basic scaling function. From the dilatation equation for the roof function (Example 3.3 where, due to the said choice of scaling function, the translation of the arguments on the right hand side by one position to the left needs to be performed), it follows that

$$\varphi(x - k) = \frac{1}{2}\,\varphi(2(x-k) - 1) + \varphi(2(x-k)) + \frac{1}{2}\,\varphi(2(x-k) + 1).$$

Taking the above notes into consideration, based on (32) and (33), we have the details as

$$\begin{aligned}
A^{(-1)}(x) - A^{(0)}(x) &= \sum_k a_{-1,k}\, \varphi(2x - k) - \sum_k a_{0,k}\, \varphi(x - k) \\
&= \sum_k a_{-1,2k+1}\, \varphi(2x - (2k+1)) + \sum_k a_{-1,2k}\, \varphi(2x - 2k) \\
&\quad - \sum_k a_{-1,2k} \left(\frac{1}{2}\,\varphi(2x - 2k - 1) + \varphi(2x - 2k) + \frac{1}{2}\,\varphi(2x - 2k + 1) \right) \\
&= \sum_k \left(a_{-1,2k+1} - \frac{1}{2}\,(a_{-1,2k} + a_{-1,2k+2}) \right) \varphi(2x - 2k - 1) \\
&= \sum_k b_{0,k}\, \varphi(2x - 2k - 1).
\end{aligned}$$

By comparing this decomposition with (33) we conclude that the wavelet obtained using the lifting algorithm without updating equals the scaling function on the previous, finer resolution level $\psi(x) = \varphi(2x-1)$. Its moment of the order zero is not equal to zero, because it is defined by the scaling function (formula (3.13)).

Updating can be performed so that the new wavelet has two vanishing moments,

$$\tag{34} \begin{aligned} \Psi(x) &= \psi(x) - \frac{1}{4}\,\varphi(x) - \frac{1}{4}\,\varphi(x-1) \\ &= \varphi(2x-1) - \frac{1}{4}\,\varphi(x) - \frac{1}{4}\,\varphi(x-1), \end{aligned}$$

Indeed, the zero moment of the wavelet $\Psi(x)$ equals zero,

$$\begin{aligned} \int \Psi(x)\,dx &= \int \varphi(2x-1)\,dx - \frac{1}{4}\int \varphi(x)\,dx - \frac{1}{4}\int \varphi(x-1)\,dx \\ &= \frac{1}{2}\int \varphi(x)\,dx - \frac{1}{4}\int \varphi(x)\,dx - \frac{1}{4}\int \varphi(x)\,dx = 0. \end{aligned}$$

The first moment of the new wavelet equals zero due to the symmetry of the initial wavelet $\psi(x)$, thus the new wavelet $\Psi(x)$ as well, relative to the point $x = 1/2$, and the proven zero vanishing moment

$$\int x\,\Psi(x)\,dx = \int \Big(x - \frac{1}{2}\Big)\,\Psi(x)\,dx + \frac{1}{2}\int \Psi(x)\,dx = 0.$$

The equation for the new wavelet, based on relations (12) and (34), is

$$\begin{aligned} \Psi(x) &= \sum_k h_1(k)\,\varphi(2x-k) - \frac{1}{4}\sum_k h_0(k)\,\varphi(2x-k-2) - \frac{1}{4}\sum_k h_0(k)\,\varphi(2x-k) \\ &= \sum_k \Big(h_1(k) - \frac{1}{4}\,h_0(k-2) - \frac{1}{4}\,h_0(k)\Big)\,\varphi(2x-k), \end{aligned}$$

whence it follows that the new wavelet coefficients $\tilde{\mathbf{h}}_1$, defining the updated wavelet $\Psi(x)$, are

$$\tag{35} \tilde{h}_1(k) = h_1(k) - \frac{1}{4}\,h_0(k-2) - \frac{1}{4}\,h_0(k).$$

The scaling function is not changed during the updating process. ∎

EXAMPLE 12. Instead of the roof function, we can start from the interpolation scaling function (24). According to (28), the wavelet attached to it is the scaling function itself on the previous (higher) resolution level, as well as a wavelet arrived at using the lifting algorithm without updating in Example 11. Formula (34) defines the updated wavelet having two vanishing moments.

The wavelets in Examples 11 and 12, though they are defined by the same formulae, are different because they have been generated by different scaling functions - the roof function in Example 11 and the interpolation function of order three in Example 12.

The following table contains the new scaling functions and wavelets coefficients obtained after the application of the lifting technique (updating (35)) to the coefficients given in Example 9.

k	-4	-3	-2	-1	0	1	2	3	4	5
$16\,h_0(k)$	0	-1	0	9	16	9	0	-1	0	0
$64\,\tilde{f}_0(k)$	1	0	-8	16	46	16	-8	0	1	0
$64\,\tilde{h}_1(k)$	0	1	0	-8	-16	46	-16	-8	0	1
$16\,\tilde{f}_1(k)$	0	0	1	0	-9	16	-9	0	1	0

Figure 5.12(b) shows the graph of the wavelet (34), arrived at by the updating of the wavelet shown on Figure 5.11. The scaling function 5.12(c) is dual to the interpolation scaling function 5.12(a), while the wavelet 5.12(d) is dual to the updated wavelet. The graphs of the dual functions are not fully fitting. They are determined, just like other graphs, by the pyramidal algorithm (§5.1) with a chosen resolution. However, the dual functions are not smooth enough and peaks increase (by their absolute value) with the increase in resolution. ∎

The updated interpolation wavelet is the simplest example of a second generation wavelet. The interpolation can be defined on more complex, irregular grids, thus these wavelets are applicable to more complex geometries. Their advantage is, also, the ability to arrive at a more compact representation, because the important properties of the modeled signal can be taken into consideration during the wavelet construction. Their failing is that they are more complex for work, because the coefficients, in general, depend on the position of the point and the resolution level.

5.7 Nonstandard wavelets

The wavelets to be briefly mentioned in the following text are called nonstandard because they do not have a compact support in the time domain. Mentioned are only wavelets most often used up to now. Details on them can be found in [20].

Morlet wavelet is a Gauss function modulated by a complex parameter,

$$\psi(x) = e^{-\imath a x}\, e^{-x^2/(2\sigma)},$$

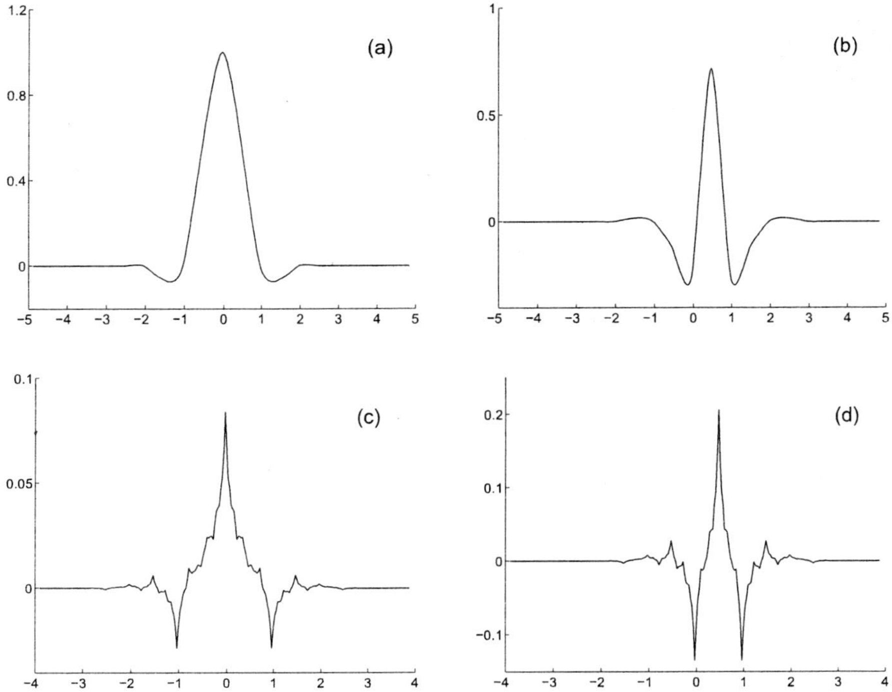

Figure 5.12: Interpolation scaling functions and updated wavelets

where a is the modulation parameter, while σ is the scaling parameter determining the window width. In order to have the first moment of this wavelet approximately zero, $\int \psi(x)\, dx \approx 0$, it should be taken that $a = \pi\,\sqrt{2/\ln 2} = 5.336$. Even though Morlet wavelet is a complex function, it is usually applied to real signals. The transformation performed by this wavelet $(f,\, \psi_{j,k})$ is represented using a module and a phase; the phase graph is especially well fit for discovering singularities.

Mexican hat is defined as the second derivative of the Gauss function

$$\omega(x) = \frac{1}{\sqrt{2\pi}\,\sigma}\, e^{-x^2/(2\sigma)},$$

yielding

$$\psi(x) = \frac{1}{\sqrt{2\pi}\,\sigma^3}\,\Big(\frac{x^2}{\sigma^2} - 1\Big)\, e^{-x^2/(2\sigma^2)}.$$

There is no scaling function.

Shannon wavelet [13] is constructed so that in the frequency domain it has a compact support. It is defined by Shannon function (see §6.1)

$$\operatorname{sinc}(\pi x) = \frac{\sin(\pi x)}{\pi x}$$

by the expression

$$\psi(x) = \frac{\sin(2\pi x) - \sin(\pi x)}{\pi x} = 2\operatorname{sinc}(2\pi x) - \operatorname{sinc}(\pi x).$$

The Fourier transform of Shannon function is the characteristic function of the interval $[-\pi, \pi)$ (a box function per frequency ω),

$$\widehat{\operatorname{sinc}}(\omega) = \aleph_{(-\pi,\pi)}(\omega) = \begin{cases} 1, & \omega \in [-\pi, \pi) \\ 0, & \omega \notin [-\pi, \pi) \end{cases}.$$

From formula (4.6) follows that the Fourier transform of Shannon function satisfies relation

$$\aleph_{(-\pi,\pi)}(\omega) = \hat{h}\left(\frac{\omega}{2}\right) \aleph_{(-\pi,\pi)}\left(\frac{\omega}{2}\right) = \hat{h}\left(\frac{\omega}{2}\right) \aleph_{(-2\pi,2\pi)}(\omega).$$

It is identity if

$$\hat{h}\left(\frac{\omega}{2}\right) = \aleph_{(-\pi,\pi)}(\omega),$$

meaning that the frequency response is

$$\hat{h}(\omega) = \aleph_{(-\pi,\pi)}(2\omega) = \aleph_{(-\pi/2,\pi/2)}(\omega).$$

By decomposing this function into a series with the form (4.5), we arrive at the following expression for the coefficients

$$h(2k) = \frac{1}{2}\delta(k), \qquad h(2k+1) = \frac{(-1)^k}{(2k+1)\pi}, \qquad k \in \mathcal{Z}.$$

It is obvious that this wavelet does not have a compact support in the time domain and it decreases slowly towards zero.

Meyer wavelet. [13] Shannon's idea was further developed by Meyer. By defining the wavelet and scaling function in the frequency domain by functions with a compact support but a greater smoothness, the scaling function and wavelet can belong to the space $\mathcal{C}^\infty$ and can decrease faster than the polynomial. Both the wavelet and the scaling function are constructed in the frequency domain using trigonometric functions, but so that their Fourier transforms have compact support,

$$\hat{\psi}(\omega) = \frac{1}{\sqrt{2\pi}}\, e^{\imath\,\omega/2} \begin{cases} \sin\left(\frac{\pi}{2}\, v\left(\frac{3}{2\pi}|\omega| - 1\right)\right) & \frac{2\pi}{3} \le |\omega| \le \frac{4\pi}{3} \\ \cos\left(\frac{\pi}{2}\, v\left(\frac{3}{4\pi}|\omega| - 1\right)\right) & \frac{4\pi}{3} \le |\omega| \le \frac{8\pi}{3} \\ 0 & |\omega| \notin \left[\frac{2\pi}{3}, \frac{8\pi}{3}\right] \end{cases}$$

$$\hat{\varphi}(\omega) = \frac{1}{\sqrt{2\pi}} \begin{cases} 1 & |\omega| \le \frac{2\pi}{3} \\ \cos\left(\frac{\pi}{2}\, v\left(\frac{3}{2\pi}|\omega| - 1\right)\right) & \frac{2\pi}{3} \le |\omega| \le \frac{4\pi}{3} \\ 0 & |\omega| > \frac{4\pi}{3} \end{cases}$$

where, for example, $v(a) = a^4(35 - 84a + 70a^2 - 20a^3)$, $a \in [0, 1]$. By changing the function v while observing certain requirements, a family of various wavelets is obtained.

The scaling function is symmetrical around point 0, while the wavelets are symmetrical around point 1/2. The scaling function and wavelet do not have compact supports, but they decrease faster than any "inverse polynomial",

$$\forall n \in N, \quad \exists c_n \qquad |\psi(x)| \le c_n \left(1 + |x|^2\right)^{-n}.$$

This wavelet is an infinitely differentiable function.

Two-dimensional wavelets. For image processing, represented by a function of two variables $F(x, y)$, it is necessary to use two-dimensional wavelets. Simple algorithms use a product of two one-dimensional scaling functions as a two-dimensional scaling function $\varphi(x, y) = \varphi(x)\,\varphi(y)$. Three two-dimensional wavelets are attached to it: the product of the scaling function and the wavelet $\psi_1(x, y) = \varphi(x)\,\psi(y)$ for representing vertical details, the product of the wavelet and the scaling function $\psi_2(x, y) = \psi(x)\,\varphi(y)$ for representing horizontal details, and the product of the two one-dimensional wavelets $\psi_3(x, y) = \psi(x)\,\psi(y)$ for representing diagonal details. The orthogonality of the two-dimensional basis thus defined is obvious if the one-dimensional wavelet basis is orthogonal. Pure two-dimensional wavelets can also be constructed, but we shall not deal with that in this book.

Which functions can be wavelets? We have seen from previous examples that the idea of wavelets, given in §2.4, can be generalized. If the function $\psi(x)$ is continuous, has moments equal to zero and decreases quickly to zero by its modulo when $x \to \infty$, or is zero outside of a finite interval, it could be a wavelet. The function $\psi(x)$ is called a wavelet if the family of dilatations and translations of that function enable all the functions with a finite energy to be reconstructed using details at all scales. The existence of an appropriate scaling function $\varphi(x)$ is not necessary. There are wavelets with no scaling function attached, like Morlet wavelet. The properties important for the choice and construction of wavelets are:

- *Compact support* of the wavelet $\psi(x)$ and the scaling function $\varphi(x)$, or their Fourier transforms $\hat{\psi}(\omega)$ and $\hat{\varphi}(\omega)$ is one of the most important wavelet properties. It may be weakened by the requirement that these functions decrease fast towards zero at infinity, either by time or frequency. This property determines the time and frequency localization. A wavelet with a compact support has a finite set of coefficients attached, thus the sums in the pyramidal algorithm (formula (3.32) and (3.33)) are finite.

- *Orthogonality* (or biorthogonality) is a desired property. The consequence of this property is the equality of the $\mathcal{L}_2$ norms of the function and the sequence of its coefficients in the decomposition by wavelets (Parseval equality (2.8)). Furthermore, the orthogonal wavelet transform is unitary, meaning that it is numerically stable. The computational algorithm is efficient as it is fast and does not consume a lot of memory. In an orthogonal multiresolution analysis the projection operators for various subspaces yield optimal approximations in the sense of the $\mathcal{L}_2$ norm.

- *Symmetry.* If the scaling function and the wavelet are symmetrical, the filters in a general case have a linear phase. If they are not symmetrical, phase distortions may appear, which is especially undesired in sound signal processing. This property is also useful for the processing of images, as two-dimensional signals. The symmetry excludes orthogonality, except in the case of the trivial Haar wavelet.

- *Regularity* is important for the smoothness of the reconstructed function, signal or image. Likewise, a greater smoothness yields better frequency localization. A smoothness of the basis functions is desired in numerical analysis, especially where derivatives are used.

- *The number of vanishing moments* of the wavelet and, if it exists, of the scaling function, is important for compression consisting of neglecting small coefficients.

- *The number of vanishing moments of the dual wavelet* determines, with smooth functions, the speed of convergence of the wavelet approximation.

- *Rational coefficients.* In computer implementations it is useful to make the scaling function and wavelet coefficients rational numbers. The algorithms are even more efficient if the coefficients are dyadic rational numbers, because in computers multiplication by powers of two is performed by shifting bits, which is an extremely fast operation.

- *Analytical expressions* do not always exist for wavelets and scaling functions, but sometimes it is desirable to have them.

- *Interpolation.* If the scaling function satisfies conditions $\varphi(k) = \delta(k)$, $k \in \mathcal{Z}$, then it is trivial to find a function from the space $\mathcal{V}_j$ interpolating the data given on a grid with a step of 2^j, because the coefficients are equal to the data.

A brief summary of the specified properties for the analyzed wavelets is given in the following table.

property	**Haar**	**DbN**	**SymN**	**CoifN**	**bior**	**inpol**	**Morl**	**Mexh**	**Meyr**
compact support	*	*	*	*	*	*			
orthogonality	*	*	*	*					*
biorthogonal.	*	*	*	*	*	*			*
symmetry	*				*	*	*	*	*
asymmetry		*							
almost symmetry			*	*					
infinite regularity							*	*	*
arbitrary regularity		*	*	*	*	*			
ψ vanish. moments		*	*	*	*				
φ vanish. moments				*					
φ exists	*	*	*	*	*	*			*
analytical expression	*				only splines		*	*	
interpolation	*					*			
continuous transform	*	*	*	*	*	*	*	*	*
discrete transform	*	*	*	*	*	*			*
fast algorithm	*	*	*	*	*	*			

Table: Summary overview of the wavelet families and their properties

It is important to stress that

> *Constructing a wavelet that has all of these properties is not possible.*

Thus various wavelet families have been constructed, depending on which properties are more important to the user.

6

Analogy with filters

We have already emphasized in the introduction that consistent wavelet theory started to develop when Stephane Mallat and Ingrid Daubechies connected approximation theory with signal processing. Still now applications that concerned signals and images are dominant.

6.1 Signal

In §2.2 a signal was defined as a function describing a physical quantity that is given for discrete argument values (a discrete signal). It represents a series of numbers obtained using an appropriate device. Unlike this, let's call it one-dimensional signal, a two-dimensional signal is called an *image*. Human speech, music, seismic or engine vibrations, financial data, medical imaging, fingerprints are only few examples of signals that need to be described efficiently, analyzed, cleaned up from noise, coded, compressed, reconstructed, simplified, modeled, separated or located. Thus the basic tasks of the scientific discipline called *signal processing* are: analysis and diagnostics, coding, quantization and compression, transfer and storage, synthesis and reconstruction. Up to about twenty years ago the main tool for signal processing was Fourier analysis.

Forming a discrete signal, i.e. sampling, is a central part of signal processing, because it is a process of discretizing a continuous quantity. The independent variable will be marked t, having in mind that this argument represents time in most cases. The sampling period should be made sufficiently small in order to make it possible to reconstruct the limited frequency function accurately based on its given values, but not too small to get redundant information.

THEOREM 1. (SAMPLING) *If* $f(t)$ *is a continuous function with a frequency range limited by* $\Omega > 0$, *then* $f(t)$ *is uniquely defined by its samples with a frequency of* 2Ω, *i.e. the values* $f(n\pi/\Omega),\, n = 0, \pm 1, \ldots$. *The minimum sampling frequency is* $\omega_s = 2\Omega$, *while the maximum allowed sampling period is*

$T = \pi/\Omega$. *The function* $f(t)$ *can be reconstructed using the interpolation formula*

$$(1) \qquad f(t) = \sum_{n=-\infty}^{\infty} f(nT)\, sinc_T(t - nT), \qquad sinc_T(t) = \frac{\sin(\pi t/T)}{\pi t/T}.$$

Proof can be found in [20]. ∎

In other words, a function which is continuous in time can be fully reconstructed based on its samples if the sampling frequency is at least twice the highest frequency in the function spectrum. The *frequency range* of the function $f(t)$ is the domain of its Fourier transform $\hat{f}(\omega)$. Function

$$\mathrm{sinc}(t) = \frac{\sin \pi t}{\pi t}$$

is called *Shannon function* (see §5.7) and it represents the inverse Fourier transform of the frequency characteristic function of the interval $[-\pi, \pi]$,

$$\aleph_{(-\pi,\pi)}(\omega) = \begin{cases} 1, & \omega \in [-\pi, \pi) \\ 0, & \omega \notin [-\pi, \pi) \end{cases}$$

because

$$\mathrm{sinc}(t) = \frac{1}{2\pi} \int_{-\infty}^{\infty} \aleph_{(-\pi,\pi)}(\omega)\, e^{\imath\omega t}\, d\omega = \frac{1}{2\pi} \int_{-\pi}^{\pi} e^{\imath\omega t}\, d\omega = \frac{\sin \pi t}{\pi t}.$$

It should be noted that $\mathrm{sinc}_T(nT) = \delta(n)$, i.e. that this function has the interpolation property, because it is equal to one for $t = 0$ and equal to zero in multiples of T different from zero.

The ratio

$$(2) \qquad \frac{1}{T} = \frac{\Omega}{\pi} \quad \text{is} \quad \textit{Nyquist speed}$$

and it defines the sampling period T.

EXAMPLE 1. Figure 6.1 shows various samples of the function $\cos \pi t$, with a frequency range limited by $\Omega = \pi$. Figure 6.1(a) represents the sample formed with the sampling period $\Delta t = T = 1$, i.e. with the Nyquist speed equals 1. Signal $x(n) = \cos n\pi$, $n = 0, \pm 1, \ldots,$ (black spots) allows perfect reconstruction of the continuous function by formula (1).

Figure 6.1(b) represents sampling with a period half as short $\Delta t = 1/2 < T$, i.e. with a speed $1/\Delta t = 2$ that is twice greater than Nyquist's. The matching signal is $x(n) = \cos{(n\pi)/2}$, $n = 0, \pm 1, \ldots,$ and it has too many data (redundancy).

Figure 6.1(c) represents sampling performed with a period $\Delta t = 3/2 > T$, i.e. with a speed $1/\Delta t = 2/3$ that is lower than Nyquist's. Signal $x(n) = \cos{(3n\pi)/2}$, $n = 0, \pm 1, \ldots,$ makes it unclear which function it represents, whether it is our

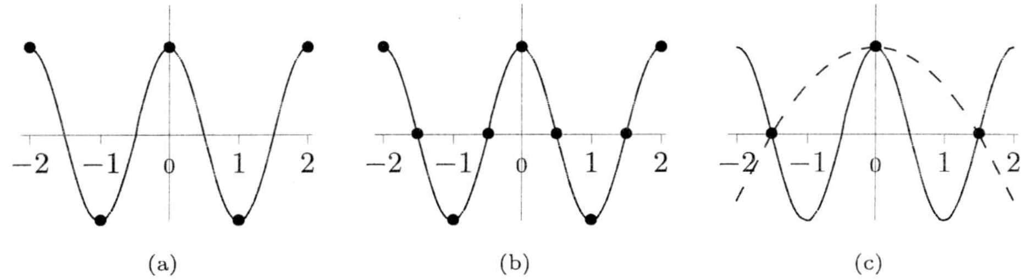

Figure 6.1: Various samples of the function $\cos \pi t$

function $\cos \pi t$ (drawn using a full line) or the function $\cos(\pi/3)\,t$, with a frequency range of $\pi/3$ (drawn using a dashed line). This phenomenon is called *aliasing*. Signal $\mathbf{x}$ does not give full information for the reconstruct of our function in this case. ∎

When $\hat{f}(\omega) \equiv 0$ for $|\omega| > \pi$, i.e. the frequency of the function $f(t)$ satisfies the condition $|\omega| \leq \pi = \Omega$ it follows that optimal sampling period is $T = 1$. The function $f(t)$ can be exactly reconstructed using its samples $f(n)$ by the interpolation formula (1),

$$f(t) = \sum_{n=-\infty}^{\infty} f(n)\, \frac{\sin \pi(t-n)}{\pi(t-n)}. \tag{3}$$

The relation between signal $f(n)$ and Fourier transform $\hat{f}(\omega)$ of function $f(t)$ is given by the following theorem.

THEOREM 2. (POISSON SUMMATION FORMULA) *For a function $f(t)$ with sufficient smoothness and decay relation between the function and it's Fourier transform is*

$$\sum_{n=-\infty}^{\infty} f(t-nT) = \frac{1}{T} \sum_{k=-\infty}^{\infty} \hat{f}\Big(\frac{2\pi k}{T}\Big)\, e^{\imath 2\pi k t/T}.$$

For $T = 1$ and $t = 0$ it gives the expression

$$\sum_{n=-\infty}^{\infty} f(n) = \sum_{k=-\infty}^{\infty} \hat{f}(2\pi k).$$

Proof can be found in [49]. ∎

6.2 Filter

It was mentioned in the Example 3.6 that Ingrid Daubechies used coefficients of an orthogonal filter as the coefficients of the dilatation equation (3.11) to obtain an orthonormal function basis. Since all the properties of a scaling function $\varphi(x)$ and a wavelet $\psi(x)$, such as their compact supports, orthogonality, smoothness and vanishing moments, stem from the properties of dilatation and wavelet equation coefficients (see §4.4), we shall analyze these properties from the aspect of digital filters. By analyzing filters we shall arrive at the conditions to be met by the coefficients of the dilatation equation in order for the wavelets to have the desired properties.

Filter is used to separate a frequency group from a signal, i.e. to separate all the components with frequencies in a given range. It is determined by the signal $\mathbf{h} = \{h(n)\}$, and acts on the input signal $\mathbf{x} = \{x(n)\}$ so that the output signal $\mathbf{y} = \{y(n)\}$ is a convolution of the signals $\mathbf{h}$ and $\mathbf{x}$ (definition 2.2),

$$(4) \qquad \mathbf{y} = \mathbf{h} * \mathbf{x}, \qquad y(n) = \sum_k h(k)\, x(n-k).$$

In signal processing so-called *causal* filters, where $h(k) = 0$ for $k < 0$, are usually used. This means that the output cannot depend on future input, because otherwise in the component $y(n)$ the addend $h(k)\, x(n+|k|)$ would appear for $k < 0$. The filter matrix F of the causal filter (formula 2.31) is a lower-triangular matrix. Mathematically, filter is a linear operator invariant in time.

Important characterization of the filter $\mathbf{h}$ is the *filter frequency response* $\hat{h}(\omega)$, already defined by formula (4.5),

$$(5) \qquad \hat{h}(\omega) = \sum_n h(n)\, e^{-\imath n\omega} \quad \text{or} \quad H(z) = \sum_n h(n)\, z^{-n}, \qquad z = e^{\imath\omega}.$$

Let us now explain its name. It represents the Fourier transform of the filter response to the unit impulse $\mathbf{x} = \{\dots, 0, 1, 0, \dots\}$ at zero time ($x(0) = 1$, while $x(n) = 0,\ n \neq 0$). Since $\hat{x}(\omega) = x(0) = 1$, it follows from formula (2.33) that $\hat{y}(\omega) = \widehat{(\mathbf{h} * \mathbf{x})}(\omega) = \hat{h}(\omega)$. The output signal $\mathbf{y}$ is the filter $\mathbf{h}$ itself. We say that the filter coefficients represent an *impulse response*, i.e. the filter response to a unit impulse.

We shall give some examples of filters.

EXAMPLE 2. If the output signal $\mathbf{y}$ and the input signal $\mathbf{x}$ are connected by a time invariant linear difference equation

$$\sum_{k=0}^{N} a_k\, y(n-k) = \sum_{k=0}^{M} b_k\, x(n-k),$$

by using the z-transform (2.34) and the delay property we find the frequency response of the filter as the ratio of the z-transformations of the input and output signals,

(6) $$H(z) = \frac{Y(z)}{X(z)} = \frac{\sum_{k=0}^{M} b_k\, z^{-k}}{\sum_{k=0}^{N} a_k\, z^{-k}},$$

because

$$\sum_n \left(\sum_k a_k\, y(n-k)\right) z^{-n} = \sum_n \left(\sum_k b_k\, x(n-k)\right) z^{-n}$$

$$\sum_k a_k \left(\sum_n y(n-k)\, z^{-n}\right) = \sum_k b_k \left(\sum_n x(n-k)\, z^{-n}\right)$$

$$Y(z) \sum_k a_k\, z^{-k} = X(z) \sum_k b_k\, z^{-k}.$$

The causal filter $h(n) = 0$ for $n < 0$, with a rational function of the frequency response, is stable if and only if all poles are within a unit circle (their modules are less than one). ∎

An *FIR (finite impulse response) filter* is one where the frequency response is given by a polynomial ($N = 0$ in expression (6)). The output is dependent solely on the input. An *IIR (infinite impulse response) filter* is one where the frequency response is given by a rational function ($1 \le N < \infty$ in expression (6)). This means that the current output depends on the previous outputs as well.

EXAMPLE 3. A *delay filter by k*, $y(n) = x(n-k)$ is a simple causal, FIR filter defined by the coefficients $h(k) = 1$ and $h(n) = 0,\, n \neq k$. The filter matrix for $k = 1$ is equal to

$$F = \begin{pmatrix} \cdot & \cdot & \cdot & \cdot & \cdot & \cdot \\ \cdot & 0 & 0 & 0 & 0 & \cdot \\ \cdot & 1 & 0 & 0 & 0 & \cdot \\ \cdot & 0 & 1 & 0 & 0 & \cdot \\ \cdot & 0 & 0 & 1 & 0 & \cdot \\ \cdot & \cdot & \cdot & \cdot & \cdot & \cdot \end{pmatrix}.$$

The z-transform of the output signal is

(7) $$Y(z) = H(z)\, X(z) = z^{-k} X(z) = \sum_n x(n)\, z^{-(n+k)} = \sum_n x(n-k)\, z^{-n}.$$

∎

EXAMPLE 4. The *averaging filter* is another simple causal FIR filter. It determines the output signal so that its elements are the mean values of the two subsequent elements of the input signal $\mathbf{x}$,

$$y(n) = \frac{1}{2}\,x(n) + \frac{1}{2}\,x(n-1), \qquad n = \cdots -1,\,0,\,1,\ldots. \tag{8}$$

The nonzero filter coefficients are $h(0) = h(1) = 1/2$. If we mark the infinitely dimensional filter matrix (2.31) as F_0, formulae (8) can be represented in a matrix form as

$$\mathbf{y} = F_0\,\mathbf{x}, \quad \text{i.e.} \quad \begin{pmatrix} \cdot \\ y(-1) \\ y(0) \\ y(1) \\ \cdot \end{pmatrix} = \begin{pmatrix} \cdot & & & & 0 \\ 1/2 & 1/2 & & & \\ & 1/2 & 1/2 & & \\ & & 1/2 & 1/2 & \\ 0 & & & \cdot & \cdot \end{pmatrix} \begin{pmatrix} \cdot \\ x(-1) \\ x(0) \\ x(1) \\ \cdot \end{pmatrix}. \tag{9}$$

The frequency response of the averaging filter is, according to (5), equal to

$$\hat{h}_0(\omega) = \frac{1}{2}\,e^0 + \frac{1}{2}\,e^{-\imath\omega} = \frac{1}{2}\big(1 + e^{-\imath\omega}\big) = \cos\frac{\omega}{2}\,e^{-\imath\omega/2}. \tag{10}$$

If we apply this filter to the signal $\mathbf{x}$ containing only one frequency ω,

$$x(n) = e^{\imath n\omega}, \qquad -\infty < n < \infty,$$

any component of the output signal $\mathbf{y}$ is the product of the frequency response (depends on ω) and the equal indexed component of the input signal $\mathbf{x}$

$$y(n) = \frac{1}{2}\,e^{\imath n\omega} + \frac{1}{2}\,e^{\imath(n-1)\omega} = \frac{1}{2}\big(1 + e^{-\imath\omega}\big)e^{\imath n\omega} = \hat{h}_0(\omega)\,x(n). \tag{11}$$

For the choice of $\omega = 0$ the input is the constant signal $\mathbf{x}_l = \{\ldots, 1,\,1,\,1,\ldots\}$, while the frequency response of the filter is $\hat{h}_0(0) = 1$. Based on (11) we conclude that the averaging filter does not change a constant signal. For low frequencies, close to $\omega = 0$, the input signal will not be changed much because $\hat{h}_0(\omega) \approx 1$. Unlike that, if we chose $\omega = \pi$, the input signal oscillates, $\mathbf{x}_h = \{\ldots, 1,\,-1,\,1,\,-1,\,1,\ldots\}$, and the frequency response of the filter is $\hat{h}_0(\pi) = 0$. This means that with the averaging filter the maximum frequency is completely dampened, as all components of the output signal are zero.

The averaging filter belongs to the *lowpass filter* group, because it does not change the low frequencies at all or changes them very little, while the high frequencies are dampened a lot or entirely. These filters separate the low frequency harmonics from a signal, like scaling functions approximate smooth function components. Lowpass filters in signal theory correspond scaling functions in wavelet theory. The box function is a continuous analogue of the averaging filter (Example

4.3). The coefficients of the dilatation equation that has the box function as a solution are just scaled averaging filter coefficients $h(k)$. Both the averaging filter and the box function smooth out the input - convolution with a box function averages in continuous time (Example 3.2),

$$(\varphi * x)(t) = \int_{-\infty}^{\infty} \varphi(t-s)x(s)\,ds = \int_{t-1}^{t} x(s)\,ds = \text{ mean value of } x(t),$$

just as the averaging filter does it in discrete time

$$\mathbf{h} * \mathbf{x} = \Big(\ldots, \frac{x(0)+x(-1)}{2}, \frac{x(1)+x(0)}{2}, \ldots\Big)$$

∎

EXAMPLE 5. The *differing filter* determines the output signal so that its elements are differences of two adjacent elements of the input signal $\mathbf{x}$,

$$y(n) = \frac{1}{2}\,x(n) - \frac{1}{2}\,x(n-1), \qquad n = \cdots -1, 0, 1, \ldots. \tag{12}$$

If the differing filter matrix (2.31) is marked by F_1, the convolution (12) can be represented in a matrix form as

$$\mathbf{y} = F_1\,\mathbf{x} \quad \text{i.e.} \quad \begin{pmatrix} \cdot \\ y(-1) \\ y(0) \\ y(1) \\ \cdot \end{pmatrix} = \begin{pmatrix} \cdot & & & & 0 \\ -1/2 & 1/2 & & & \\ & -1/2 & 1/2 & & \\ & & -1/2 & 1/2 & \\ 0 & & & \cdot & \cdot \end{pmatrix} \begin{pmatrix} \cdot \\ x(-1) \\ x(0) \\ x(1) \\ \cdot \end{pmatrix}. \tag{13}$$

The frequency response of the differing filter equals

$$\hat{h}_1(\omega) = \frac{1}{2}\,e^0 - \frac{1}{2}\,e^{-\imath\omega} = \frac{1}{2}\big(1 - e^{-\imath\omega}\big) = \sin\frac{\omega}{2}\,\imath\, e^{-\imath\omega/2}. \tag{14}$$

As we did in Example 4, we shall analyze the action of this filter on the input signal containing only one frequency, $x(n) = e^{\imath n\omega}$. The element of the output signal $\mathbf{y}$ is

$$y(n) = \frac{1}{2}\,e^{\imath n\omega} - \frac{1}{2}\,e^{\imath(n-1)\omega} = \frac{1}{2}\big(1 - e^{-\imath\omega}\big)\,e^{\imath n\omega} = \hat{h}_1(\omega)\,x(n). \tag{15}$$

Since $\hat{h}_1(0) = 0$ and $\hat{h}_1(\pi) = 1$, the filter cancels out the low frequency signal $\mathbf{x}_l$ (for $\omega = 0$) while the high frequency signal $\mathbf{x}_h$ (for $\omega = \pi$) is unchanged. Since the differing filter strongly or completely dampens low frequencies, while the high frequencies are left completely or mostly unchanged, the filter belongs to the *highpass filter* group. These filters are used to separate the high frequency harmonics from a signal. Highpass filters match wavelets. Haar's wavelet is a continuous analogue of the differing filter – it picks out changes. ∎

Generally, for an arbitrary filter and the input signal $x(n) = e^{\imath n\omega}$ containing only one frequency, any component of an output signal is equal to the product of the filter frequency response and the equal indexed component of an input signal,

$$y(n) = \sum_{k=0}^{\infty} h(k)\, x(n-k) = \sum_{k=0}^{\infty} h(k)\, e^{\imath (n-k)\omega}$$
$$= \left(\sum_{k=0}^{\infty} h(k)\, e^{-\imath k\omega} \right) x(n) = \hat{h}(\omega)\, x(n).$$

An ideal lowpass filter is one where the frequency response is (Figure 6.2, a)

(16) $$\hat{h}_0(\omega) = \begin{cases} 1, & 0 \leq |\omega| < \pi/2 \\ 0, & \pi/2 < |\omega| \leq \pi \end{cases}.$$

An ideal highpass filter is one where the frequency response is (Figure 6.2, b)

(17) $$\hat{h}_1(\omega) = \begin{cases} 0, & 0 \leq |\omega| < \pi/2 \\ 1, & \pi/2 < |\omega| \leq \pi \end{cases}.$$

We treat only the basic interval $-\pi \leq \omega \leq \pi$ because the frequency response (5) is obviously a 2π-periodic function $\hat{h}(\omega + 2\pi) = \hat{h}(\omega)$.

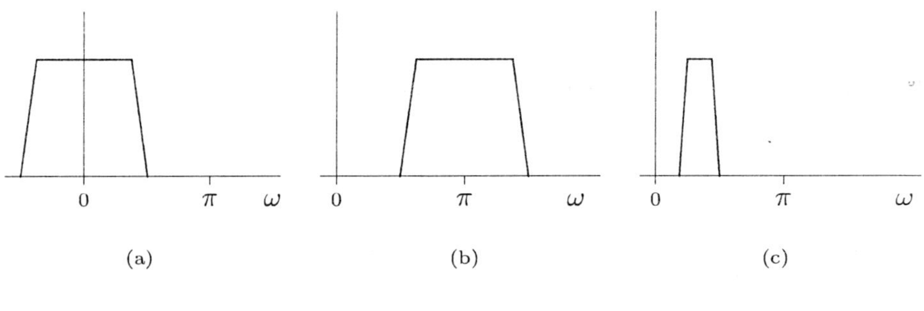

Figure 6.2: Ideal filters: lowpass (a), highpass (b) and bandpass (c)

The averaging filter and the differing filter are not invertible filters, because both of them transform some signal to zero. The averaging filter transforms to zero oscillating signal $\mathbf{x}_h$ (formula (11) for $\omega = \pi$), while the differing filter transforms to zero the constant signal $\mathbf{x}_l$) (formula (15) for $\omega = 0$). Thus not every input signal can be reconstructed based on the output signal. Expressed through the frequency response condition, the filter is not invertible if $\hat{h}(\omega) = 0$ for some ω. We see complete analogy between actions produced on signals by the described filters and on functions by Haar's scaling function (the box function) and wavelet.

6.3 Orthogonal filter bank

Although some input signals transformed by the averaging filter or the differing filter can not be reconstructed by this filter, all input signals can be reconstructed based on the output signals if the filters described are observed in pair. A highpass differing filter is attached to a lowpass averaging filter and they are *Quadrature Mirror Filters* (QMF) – they are symmetric to each other like mirror images.

A set of filters is called a *filter bank*. We distinguish an *analysis* filter bank and a *synthesis* filter bank. The former decomposes the input signal into frequency groups, while the latter reconstructs the input signal from its components (signals per frequency group). The simplest filter bank is a bank with two filters and is called a two-channel bank. A two-channel analysis bank has one lowpass filter and one highpass filter. They decompose the input signal into two frequency groups. The sub-signals thus provided can be compressed far more efficiently than the input signal, and thus transferred or stored. They can always be merged using an appropriate synthesis bank. It is not necessary to preserve all the signal components coming out of the analysis bank. Only the even components of the lowpass and highpass output are kept.

A bank containing M filters provides M signals at the output, and every M-th component of each of the output signals is kept. This means that always the total length of the output signals is equal to the length of the input signal. FIR banks of perfect reconstruction filters, where the output signal from a synthesis bank is equal to the input signal to the analysis bank, are especially interesting.

EXAMPLE 6. Let us now illustrate how the two-channel filter bank consisting of a lowpass averaging filter (Example 4) and a highpass differing filter (Example 5) works. Formulae (9) and (13), applied to input signal $\mathbf{x}$, produce the signals

$$F_0\,\mathbf{x} = \frac{1}{2}\begin{pmatrix} \vdots \\ x(-1)+x(-2) \\ x(0)+x(-1) \\ x(1)+x(0) \\ x(2)+x(1) \\ \vdots \end{pmatrix} \qquad F_1\,\mathbf{x} = \frac{1}{2}\begin{pmatrix} \vdots \\ x(-1)-x(-2) \\ x(0)-x(-1) \\ x(1)-x(0) \\ x(2)-x(1) \\ \vdots \end{pmatrix}.$$

Now we have to compress the obtained signals. The downsampling operator $(\downarrow 2)$ is used to eliminate odd and keep even signal components,

$$\mathbf{y}_0 = (\downarrow 2)\,F_0\,\mathbf{x}, \qquad \mathbf{y}_1 = (\downarrow 2)\,F_1\,\mathbf{x},$$

which produces the following output signals

$$\mathbf{y}_0 = \frac{1}{2}\begin{pmatrix} \vdots \\ x(-2)+x(-3) \\ x(0)+x(-1) \\ x(2)+x(1) \\ \vdots \end{pmatrix} \qquad \mathbf{y}_1 = \frac{1}{2}\begin{pmatrix} \vdots \\ x(-2)-x(-3) \\ x(0)-x(-1) \\ x(2)-x(1) \\ \vdots \end{pmatrix}.$$

For example, the analysis (decomposition) of the input signal $\mathbf{x}$ given in the Example 5.1 produces two output signals, the low frequency signal $\mathbf{y}_0$ and the high frequency signal $\mathbf{y}_1$,

$$\begin{array}{lcccccccc} \mathbf{x}: & \underbrace{37, \quad 35} & & \underbrace{28, \quad 28} & & \underbrace{58, \quad 18} & & \underbrace{21, \quad 15} & \\ \frac{1}{2}* & +\swarrow & \searrow - & +\swarrow & \searrow - & +\swarrow & \searrow - & +\swarrow & \searrow - \\ \mathbf{y}_0: & 36 & & 28 & & 38 & & 18 & \\ \mathbf{y}_1: & & 1 & & 0 & & 20 & & 3 \end{array}.$$

It is obvious that for these simple filters the synthesis (reconstruction) of the signal $\mathbf{x}$ can be performed by adding and subtracting the signals $\mathbf{y}_0$ and $\mathbf{y}_1$, yielding even or odd components of the input signal $\mathbf{x}$. In a general case the synthesis procedure, i.e. reconstruction of the input signal $\mathbf{x}$ is performed by use of two filters from the synthesis filter bank. Before that, by inserting zeroes in place of the odd components, the signals $\mathbf{y}_0$ and $\mathbf{y}_1$ are expanded to their full length (equal to the dimensions of the signal $\mathbf{x}$),

$$\mathbf{u}_0 = (\uparrow 2)\,\mathbf{y}_0 = \frac{1}{2}\begin{pmatrix} \vdots \\ x(-2)+x(-3) \\ 0 \\ x(0)+x(-1) \\ 0 \\ x(2)+x(1) \\ \vdots \end{pmatrix} \qquad \mathbf{u}_1 = (\uparrow 2)\,\mathbf{y}_1 = \frac{1}{2}\begin{pmatrix} \vdots \\ x(-2)-x(-3) \\ 0 \\ x(0)-x(-1) \\ 0 \\ x(2)-x(1) \\ \vdots \end{pmatrix}.$$

Decompression is done by use of the upsampling operator $(\uparrow 2)$ that is inverse to the downsampling operator $(\downarrow 2)$. These operators are defined by matrices

$$(\downarrow 2) = \begin{pmatrix} \cdot & \cdot & \cdot & \cdot & \cdot & \cdot & \cdot \\ \cdot & 1 & 0 & 0 & 0 & 0 & \cdot \\ \cdot & 0 & 0 & 1 & 0 & 0 & \cdot \\ \cdot & 0 & 0 & 0 & 0 & 1 & \cdot \\ \cdot & 0 & 0 & 0 & 0 & 0 & \cdot \\ \cdot & \cdot & \cdot & \cdot & \cdot & \cdot & \cdot \end{pmatrix}, \qquad (\uparrow 2) = (\downarrow 2)^{\top} = \begin{pmatrix} \cdot & \cdot & \cdot & \cdot & \cdot & \cdot \\ \cdot & 1 & 0 & 0 & 0 & \cdot \\ \cdot & 0 & 0 & 0 & 0 & \cdot \\ \cdot & 0 & 1 & 0 & 0 & \cdot \\ \cdot & 0 & 0 & 0 & 0 & \cdot \\ \cdot & 0 & 0 & 1 & 0 & \cdot \\ \cdot & \cdot & \cdot & \cdot & \cdot & \cdot \end{pmatrix},$$

which are orthogonal and so transposed to each other.

After decompression we perform filtering by the synthesis bank. Vectors $\mathbf{u}_0$ and $\mathbf{u}_1$ are the inputs for two synthesis filters G_0 and G_1. Nonzero elements of the filter G_0 are $g_0(0) = 1$ and $g_0(1) = 1$, and nonzero elements of the filter G_1 are $g_1(0) = -1$ and $g_1(1) = 1$. Thereby filter G_0 adds while filter G_1 subtracts two subsequent coordinates of the input signal,

$$w_0(n) = u_0(n) + u_0(n-1), \qquad w_1(n) = -u_1(n) + u_1(n-1).$$

The output signals obtained by these filters are

$$\mathbf{w}_0 = G_0\,\mathbf{u}_0 = \frac{1}{2}\begin{pmatrix} \vdots \\ x(-2)+x(-3) \\ x(-2)+x(-3) \\ x(0)+x(-1) \\ x(0)+x(-1) \\ x(2)+x(1) \\ \vdots \end{pmatrix} \qquad \mathbf{w}_1 = G_1\,\mathbf{u}_1 = \frac{1}{2}\begin{pmatrix} \vdots \\ -x(-2)+x(-3) \\ x(-2)-x(-3) \\ -x(0)+x(-1) \\ x(0)-x(-1) \\ -x(2)+x(1) \\ \vdots \end{pmatrix},$$

and their addition provides the input signal $\mathbf{x}$ with a delay, because the output $x(n-1)$ matches the input $x(n)$. ∎

To resume, the signal analysis procedure is performed in two steps: filtering and compression. They can be united by leaving out the odd rows in the filter matrix. Thus for the basic filters used in Example 6. we arrive at rectangular matrices

$$C = (\downarrow 2)\,\sqrt{2}\,F_0 = \begin{pmatrix} 1/\sqrt{2} & 1/\sqrt{2} & & & \\ & & 1/\sqrt{2} & 1/\sqrt{2} & \\ & & & & \ddots \end{pmatrix},$$

(18)

$$D = (\downarrow 2)\,\sqrt{2}\,F_1 = \begin{pmatrix} -1/\sqrt{2} & 1/\sqrt{2} & & & \\ & & -1/\sqrt{2} & 1/\sqrt{2} & \\ & & & & \ddots \end{pmatrix}.$$

Multiplying by factor $\sqrt{2}$ has been performed to make the square analysis bank matrix, obtained by merging two matrices C and D, an orthogonal matrix

$$\begin{pmatrix} C \\ D \end{pmatrix} = \frac{1}{\sqrt{2}} \begin{pmatrix} 1 & 1 & & & \\ & & 1 & 1 & \\ & & & & \cdot \\ -1 & 1 & & & \\ & & -1 & 1 & \\ & & & & \cdot \end{pmatrix}. \tag{19}$$

The analysis of the signal $\mathbf{x}$ can be written down by the following formula

$$\mathbf{y} = \begin{pmatrix} C \\ D \end{pmatrix} \mathbf{x} = \sqrt{2}\,(\downarrow 2) \begin{pmatrix} F_0 \\ F_1 \end{pmatrix} \mathbf{x}. \tag{20}$$

The reconstruction of the input signal $\mathbf{x}$ from the output signal $\mathbf{y}$ (signal synthesis) can be performed by solving the equation (20) per $\mathbf{x}$,

$$\mathbf{x} = \begin{pmatrix} C \\ D \end{pmatrix}^{-1} \mathbf{y} = \frac{1}{\sqrt{2}} \left(F_0^\top \quad F_1^\top\right) (\uparrow 2)\,\mathbf{y}, \tag{21}$$

where we use the orthogonality of the analysis bank matrix (19). Synthesis of a signal is performed in two steps – decompression and filtering. The first step is to obtain signals of the full length by inserting zeroes as signal elements; upsampling operator $(\uparrow 2)$ inserts zeroes as the odd elements of the signal (the signal length is doubled). The second step is filtering that is performed by synthesis filter bank. In this example the analysis and synthesis algorithms use the same filters because the analysis filter bank is orthogonal. The analysis and synthesis of a signal is performed by the same filter bank if it is orthogonal, i.e. the filter bank is characterized by an orthogonal matrix. The orthogonal filter bank made up of the lowpass averaging filter and the highpass differing filter is called the *Haar filter bank*. The synthesis bank matrix is transposed to the analysis bank matrix (19)

$$\begin{pmatrix} C \\ D \end{pmatrix}^{-1} = \frac{1}{\sqrt{2}} \begin{pmatrix} 1 & & -1 & \\ 1 & & 1 & \\ & 1 & & -1 \\ & 1 & & 1 \\ & \cdot & & \cdot \end{pmatrix}.$$

The lowpass filter matrix C is the same as the dilatation equation matrix M, defined in (4.12) for the box function coefficients $h(0) = h(1) = 1/2$, and

$c(k) = \sqrt{2}\, h(k)$. The elements of the highpass filter matrix D are coefficients of the wavelet equation that defines Haar wavelet (Example 3.2). Therefore matrix C will be attached to the scaling function and matrix D will be attached to the wavelet. This filter bank is called Haar filter bank and it is orthogonal, such as the basis defined by the box function (Haar function) and the attached wavelet. If the wavelet is seen as a highpass filter and the scaling function as a lowpass filter, the set of compressed wavelets along with the scaling function can be seen as the filter bank. Each of the wavelets covers a certain frequency range, while the scaling function covers the rest of the spectrum including smaller frequencies (greater scale). Discrete wavelet transform corresponds to the signal analysis, while inverse discrete wavelet transform corresponds to the signal synthesis (Figure 5.1).

The compression operation (downsampling) corresponds to a reduction in sample density, i.e. the removal of certain signal elements. For example, compression by two means leaving out every other element of the signal. The decompression operation (upsampling) corresponds to the increase of the sample density by adding new elements to the signal. Decompression by two means adding zero or an interpolated value between every two signal elements. These operations represent a multiresolution of the signals and bring filters and wavelets into relation. Namely, the output signal $\mathbf{y} = \mathbf{h} * \mathbf{x}$ has the even indexed elements

$$y(2n) = \sum_k h(k)\, x(2n-k).$$

By compression by two, i.e. leaving out the odd elements and renumbering the others, the output signal $(\downarrow 2)\,\mathbf{y}$ is obtained the n-th element of which is

$$y(2n) \quad \to \quad y(n) = \sum_k h(k)\, x(2n-k).$$

This is a two-scale relation just like the dilatation equation (3.11).

Downsampling $y(n) = x(n\,M)$ by an integer factor M (keeping every M-th element) yields the output signal whose spectrum is dilated by M,

$$\text{(22)} \qquad \hat{y}(\omega) = \frac{1}{M}\sum_{k=0}^{M-1} \hat{x}\Big(\frac{\omega - 2\pi k}{M}\Big), \qquad Y(z) = \frac{1}{M}\sum_{k=0}^{M-1} X\big(W_M^k\, z^{1/M}\big),$$

where $W_M = e^{\imath 2\pi/M}$. Upsampling by an integer factor M produces output signal with nonzero elements $y(n) = x(n/M)$ for $n = kM$, $k \in \mathcal{Z}$. It's spectrum is contracted M times,

$$\text{(23)} \qquad \hat{y}(\omega) = \hat{x}(M\,\omega), \qquad Y(z) = X(z^M).$$

We shall consider now more general case of a two-channel FIR orthogonal filter bank characterized with an orthogonal matrix. The same filters are used for analysis and synthesis and the filter bank has the property of perfect reconstruction, meaning that the output signal from the synthesis bank is equal to the input signal of the analysis bank. It will be proved that such filters satisfy double shift orthogonality conditions (3.26), representing the orthogonality conditions for the scaling function and wavelet basis as well (Theorem 3.1).

An infinitely dimensional analysis bank orthogonal matrix (analog to (19)) has the form

$$\begin{pmatrix} C \\ D \end{pmatrix} = \begin{pmatrix} \cdot & \cdot & \cdot & \cdot & \cdot & \cdot & \cdot & \cdot \\ \cdot & c(3) & c(2) & c(1) & c(0) & 0 & 0 & \cdot \\ \cdot & c(5) & c(4) & c(3) & c(2) & c(1) & c(0) & \cdot \\ \cdot & \cdot & \cdot & \cdot & \cdot & \cdot & \cdot & \cdot \\ \cdot & d(3) & d(2) & d(1) & d(0) & 0 & 0 & \cdot \\ \cdot & d(5) & d(4) & d(3) & d(2) & d(1) & d(0) & \cdot \\ \cdot & \cdot & \cdot & \cdot & \cdot & \cdot & \cdot & \cdot \end{pmatrix}. \tag{24}$$

A shift by two in rows, i.e. the deletion of every other row is a consequence of the operation of compression by two. As the analysis filter bank is used in perfect reconstruction synthesis,

$$\mathbf{y} = \begin{pmatrix} C \\ D \end{pmatrix} \mathbf{x}, \qquad \mathbf{x} = \begin{pmatrix} C \\ D \end{pmatrix}^{\top} \mathbf{y},$$

the synthesis bank matrix is transposed to the analysis bank matrix

$$\begin{pmatrix} C \\ D \end{pmatrix}^{\top} = \begin{pmatrix} C^{\top} & D^{\top} \end{pmatrix} = \begin{pmatrix} \cdot & \cdot & \cdot & \cdot & \cdot & \cdot & \cdot \\ \cdot & c(3) & c(5) & \cdot & d(3) & d(5) & \cdot \\ \cdot & c(2) & c(4) & \cdot & d(2) & d(4) & \cdot \\ \cdot & c(1) & c(3) & \cdot & d(1) & d(3) & \cdot \\ \cdot & c(0) & c(2) & \cdot & d(0) & d(2) & \cdot \\ \cdot & 0 & c(1) & \cdot & 0 & d(1) & \cdot \\ \cdot & 0 & c(0) & \cdot & 0 & d(0) & \cdot \\ \cdot & \cdot & \cdot & \cdot & \cdot & \cdot & \cdot \end{pmatrix}.$$

A shift by two in columns, i.e. the deletion of every other column, is the consequence of the operation of decompression by two. The orthogonality condition of the filter bank matrix

$$\begin{pmatrix} C \\ D \end{pmatrix} \begin{pmatrix} C^{\top} & D^{\top} \end{pmatrix} = \begin{pmatrix} C\,C^{\top} & C\,D^{\top} \\ D\,C^{\top} & D\,D^{\top} \end{pmatrix} = \begin{pmatrix} I & 0 \\ 0 & I \end{pmatrix}$$

i.e.

$$C\,C^{\top} = I, \qquad C\,D^{\top} = 0, \qquad D\,D^{\top} = I,$$

leads to the well known double shift orthogonality conditions (3.26)

$$\sum_k c(k)\, c(k-2n) = \delta(n),$$

$$\sum_k c(k)\, d(k-2n) = 0,$$

$$\sum_k d(k)\, d(k-2n) = \delta(n).$$

As it has been said filters with odd lengths cannot have this property. For example, with the length of $N = 3$, we have

$$\big(c(0), c(1), c(2)\big) \cdot \big(0, 0, c(0)\big)^{\top} = c(0)c(2) \neq 0.$$

If the filter length N is an even number, the double shift orthogonality conditions yield the correlation between the coefficients $c(k)$ of the lowpass filter and the coefficients $d(k)$ of the highpass filter,

$$d(k) = (-1)^k\, c(N-1-k), \qquad k = 0, \dots, N-1, \quad N \text{ even},$$

which is the already known relation (3.27) between dilatation and wavelet equation coefficients. The orthogonal filter bank matrix with the length $N = 4$ has the form

$$\begin{pmatrix} \cdot & \cdot & \cdot & \cdot & \cdot & \cdot & \cdot & \cdot \\ \cdot & c(3) & c(2) & c(1) & c(0) & 0 & 0 & \cdot \\ \cdot & 0 & 0 & c(3) & c(2) & c(1) & c(0) & \cdot \\ \cdot & \cdot & \cdot & \cdot & \cdot & \cdot & \cdot & \cdot \\ \cdot & -c(0) & c(1) & -c(2) & c(3) & 0 & 0 & \cdot \\ \cdot & 0 & 0 & -c(0) & c(1) & -c(2) & c(3) & \cdot \\ \cdot & \cdot & \cdot & \cdot & \cdot & \cdot & \cdot & \cdot \end{pmatrix}.$$

To resume, a two-channel FIR filter bank is a perfect reconstruction filter bank if it is orthogonal, i.e. if it satisfies CONDITION O

$$(25) \quad \sum_k c(k)\, c(k-2n) = \delta(n), \quad d(k) = (-1)^k\, c(N-1-k), \quad k = 0, \dots, N-1.$$

Orthogonality excludes coefficient symmetry. A symmetrical orthogonal FIR filter can only have two coefficients different from zero, and this is the Haar filter (Example 4). A lowpass filter with the coefficients $c(k)$ and a highpass filter with the coefficients $d(k)$, both N in length, are filters with mirror symmetry if their coefficients meet relations (25).

Filtering and compression provide two signals at the analysis output, the low frequency part $\mathbf{y}_0$ and the high frequency part $\mathbf{y}_1$, the elements of which are

$$(26) \qquad y_0(n) = \sum_k c(2n-k)\,x(k), \qquad y_1(n) = \sum_k d(2n-k)\,x(k).$$

The length of each of them is equal to half of the input signal length. Reconstruction provides a signal $\tilde{\mathbf{x}}$ of the initial length at the synthesis output, the elements of which are

$$(27) \qquad \tilde{x}(k) = \sum_n \big(c(2n-k)\,y_0(n) + d(2n-k)\,y_1(n) \big).$$

Expressions (26) are equal to the pyramid algorithm formulae (3.32) and expression (27) is equal to the inverse pyramid algorithm formula (3.33) (the opposite sign of index $2n-k$ appears only because the transpose matrix is used for the analysis step).

6.4 Daubechies filters

Now we want to see how to construct an orthogonal filter bank based on the shortest possible filter that fulfills CONDITION A_r (4.22) for given integer r. Such filter as the vector of dilatation equation coefficients will define orthonormal wavelet basis with nice approximation properties (§4.4). Ingrid Daubechies solved this problem in 1988 [19]. She constructed a family of orthogonal FIR filters with the length $2r$ and with the r-th order zero of the frequency response in the point $\omega = \pi$. The matching wavelets have a compact support on the interval $[0,\, 2r-1]$. As r increases, the filter regularity increases, and the wavelet smoothness increases as well.

Let us first see how the orthogonality condition is expressed in the frequency domain. We shall write the filter frequency response (5) using normalized coefficients $c(k) = \sqrt{2}\,h(k)$ that define orthogonal filter matrix (24),

$$\hat{c}(\omega) = \sum_k c(k)\,e^{-\imath k\omega},$$

and its z-transform

$$C(z) = \sum_k c(k)\,z^{-k}, \qquad z = e^{\imath\omega}.$$

Note the relation

$$\begin{aligned}\hat{c}(\omega+\pi) &= \sum c(k)\,e^{-\imath k(\omega+\pi)} = \sum c(k)\,\left(e^{\imath\omega}\,e^{\imath\pi}\right)^{-k} \\ &= \sum c(k)\,\left(-e^{\imath\omega}\right)^{-k} = \sum c(k)\,(-z)^{-k} = C(-z).\end{aligned}$$

The power spectral response $P(z)$ is a square of the modulo of the filter frequency response and can be expressed as follows

$$
\begin{aligned}
P(z) &= |C(z)|^2 = C(z)\,\overline{C(z)} = C(z)\,C(z^{-1}) \\
&= \left(\sum_{k=0}^{N-1} c(k)\,z^{-k}\right)\left(\sum_{l=0}^{N-1} c(l)\,z^{l}\right) = \sum_{k=0}^{N-1}\sum_{l=0}^{N-1} c(k)\,c(l)\,z^{-(k-l)} \qquad (28)\\
&= \sum_{n=-N+1}^{N-1}\left(\sum_{k=0}^{N-1} c(k)\,c(k-n)\right) z^{-n} = \sum_{n=-N+1}^{N-1} p(n)\,z^{-n},
\end{aligned}
$$

because $|z| = 1$ and $\overline{z} = z^{-1}$.

If the filter is orthogonal the coefficients of $P(z)$ are, based on (25), equal to

$$
(29) \qquad p(2m) = \sum_k c(k)\,c(k-2m) = \begin{cases} 1 & \text{if } m = 0 \\ 0 & \text{if } m \neq 0 \end{cases},
$$

meaning that $P(z)$ is a polynomial (with negative powers by z also) where, other than the constant 1, there are only odd powers by z. The orthogonality condition in the frequency domain is, thus,

$$
(30) \qquad |\hat{c}(\omega)|^2 + |\hat{c}(\omega+\pi)|^2 = 2, \qquad \text{or} \qquad |C(z)|^2 + |C(-z)|^2 = 2,
$$

i.e.

$$
(31) \qquad \hat{p}(\omega) + \hat{p}(\omega+\pi) = 2, \qquad \text{or} \qquad P(z) + P(-z) = 2.
$$

The function $\hat{p}(\omega) = |\hat{c}(\omega)|^2$ is equal to its conjugated-complex function, because it is real and non-negative, which means, based on (28), that $p(n) = p(-n)$ as well. Taking (29) in consideration also $P(z)$ can be represented by cosine functions

$$
\begin{aligned}
P(z) &= 1 + \sum_{k=1}^{N/2} p(2k-1)\left(z^{-(2k-1)} + z^{2k-1}\right) \qquad (32)\\
&= 1 + 2\sum_{k=1}^{N/2} p(2k-1)\,\cos\left((2k-1)\omega\right).
\end{aligned}
$$

It means that the power spectral response of the orthogonal filter is a polynomial by the odd powers of the argument $\cos\omega$.

The coefficients $p(n)$ are defined by the autocorrelation of the filter $\mathbf{c}$,

$$
p(n) = \sum_k c(k)\,c(k-n), \qquad \mathbf{p} = \mathbf{c} * \mathbf{c}^\top,
$$

representing a convolution of the signal $\mathbf{c} = (c(0),\, c(1),\, c(2), \ldots)$ with its time reverse $\mathbf{c}^T = (\ldots, c(2),\, c(1),\, c(0))$. The substitution of $(-n)$ with n in the last expression does not change $p(n)$ because $p(n) = p(-n)$. The autocorrelation filter $P(z)$ which satisfies the condition (31) is called a *halfband filter*. A highpass response $\hat{d}(\omega)$ similarly leads to the autocorrelation filter $\hat{p}_1(\omega) = |\hat{d}(\omega)|^2$.

EXAMPLE 7. The simplest form of the power spectral response (32) which satisfies the basic condition $P(z) \geq 0$ is

$$P(z) = 1 + \frac{z^{-1} + z}{2}, \qquad \text{i.e.} \qquad \hat{p}(\omega) = 1 + \cos\omega.$$

Since $\hat{p}(0) = 2$ and $\hat{p}(\pi) = 0$, condition (31) has been met, thus this filter is a halfband filter. From the factorization

$$\hat{p}(\omega) = \frac{1}{2}\,(1 + e^{-\imath\omega})\,(1 + e^{\imath\omega}) = |\hat{c}(\omega)|^2,$$

we obtain the frequency response of the orthogonal filter $\mathbf{c}$,

$$\hat{c}(\omega) = \frac{1}{\sqrt{2}}\,(1 + e^{-\imath\omega}).$$

The coefficients $c(0) = 1/\sqrt{2}$, $c(1) = 1/\sqrt{2}$, i.e. $h(0) = 1/2$, $h(1) = 1/2$ are the coefficients of the well-known Haar averaging filter (Example 4) that generates the orthogonal box function wavelet basis. ∎

EXAMPLE 8. The roof function (linear spline) defined in the Example 3.3, is the solution of the dilatation equation with the coefficients $c(0) = 1/(2\sqrt{2})$, $c(1) = 1/\sqrt{2}$, $c(2) = 1/(2\sqrt{2})$. The frequency response of the coefficient filter $\mathbf{c}$,

$$\hat{c}(\omega) = \frac{1}{\sqrt{2}}\left(\frac{1}{2} + e^{-\imath\omega} + \frac{1}{2}e^{-2\imath\omega}\right) = \frac{1}{\sqrt{2}}\left(\frac{1 + e^{-\imath\omega}}{\sqrt{2}}\right)^2,$$

is scaled by $\frac{1}{\sqrt{2}}$ the square of the frequency response of the Haar filter analyzed in Example 7. The power spectral response

$$\hat{p}(\omega) = |\hat{c}(\omega)|^2 = \frac{1}{2}\,(1 + \cos\omega)^2,$$

does not meet condition (31), i.e. it is not a halfband filter, so the lowpass filter $\hat{c}(\omega)$ does not belong to the orthogonal filter bank. In wavelet theory we have already concluded that the roof function is not orthogonal to its translations. ∎

EXAMPLE 9. If we multiply the power spectral response from the Example 8 with the proper factor to meet the condition (31),

$$\hat{p}(\omega) = (1+\cos\omega)^2\,(1-\frac{1}{2}\cos\omega),$$

we shall get the Daubechies filter Db2. Note that the power spectral response has a double zero at the point $\omega=\pi$. By multiplication with the second factor (which is also positive), the even powers by $\cos\omega$ are canceled out,

$$\hat{p}(\omega) = 1+\frac{3}{2}\cos\omega-\frac{1}{2}\cos^3\omega,$$

or, for $\cos\omega=\frac{1}{2}\left(z+z^{-1}\right)$ we get a halfband filter

$$P(z) = -\frac{1}{16}\,z^3+\frac{9}{16}\,z+1+\frac{9}{16}\,z^{-1}-\frac{1}{16}\,z^{-3}.$$

Its factor a lowpass orthogonal filter $C(z)$ (relation (28)) has to be determined. In the general case this is not a simple task. We already know the part of $C(z)$ originating from the square factor, so we shall determine the part originating with the second factor

$$\hat{q}(\omega) = 1-\frac{1}{2}\cos\omega, \qquad \text{or} \qquad Q(z) = 1-\frac{z^{-1}+z}{4}.$$

We are to find the polynomial by z that, multiplied with a polynomial conjugated to it gives $Q(z)$. The condition

$$1-\frac{1}{4}\left(z^{-1}+z\right) = \left(b(0)+b(1)z^{-1}\right)\,\left(b(0)+b(1)z\right)$$

yields

$$\begin{aligned} b(0)^2+b(1)^2 &= 1, \\ b(0)\,b(1) &= -\frac{1}{4} \end{aligned} \qquad \longrightarrow \qquad b(0)=\frac{1+\sqrt{3}}{\sqrt{8}}, \qquad b(1)=\frac{1-\sqrt{3}}{\sqrt{8}}$$

Thus, using also the result of the previous example,

$$C(z) = \frac{1}{2}(1+z^{-1})^2\,\left(b(0)+b(1)z^{-1}\right),$$

we get the frequency response of the famous *Daubechies* Db2 filter (named after its creator, Ingrid Daubechies)

$$C(z) = \frac{1}{4\sqrt{2}}\left((1+\sqrt{3})+(3+\sqrt{3})z^{-1}+(3-\sqrt{3})z^{-2}+(1-\sqrt{3})z^{-3}\right). \tag{33}$$

The scaling function was defined just by these coefficients in Example 3.6. The frequency response $C(z)$ satisfies the CONDITION A_2 as it has a double zero for $z=-1$, i.e. $\hat{c}(\omega)$ has a double zero for $\omega=\pi$. This means that the function $C(z)$ is flat in the point π and that the wavelet has two vanishing moments (§4.4). ∎

Ingrid Daubechies generalized the idea described in Example 9 thus arriving at *Daubechies maxflat filters* Dbr. Their construction is based on two key properties:

1. Filters (and wavelets) are orthogonal.
2. Power spectral responses are maxflat for $\omega = \pi$ (wavelet bases have approximation properties A_r).

The requirement that the frequency response $\hat{c}(\omega)$ should have a zero of the order r in the point π means that it has a factor $(1 + e^{-\imath\omega})^r$,

$$\hat{c}(\omega) = \left(\frac{1 + e^{-\imath\omega}}{2}\right)^r \hat{q}(\omega). \tag{34}$$

This factor is normed by 2 so that in the point $\omega = 0$ it has the value 1. The second factor $\hat{q}(\omega)$ is a polynomial by $z^{-1} = e^{-\imath\omega}$ of the degree $r - 1$ because $\hat{c}(\omega)$ is a polynomial by z^{-1} of the degree $2r - 1$. Since CONDITIONS A_r are met by writing $\hat{c}(\omega)$ in the form of (34), the r coefficients of the polynomial $\hat{q}(\omega)$ are determined based on CONDITION O. Daubechies obtained that the power spectral response of her Dbr filter is equal to

$$\hat{p}(\omega) = |\hat{c}(\omega)|^2 = 2\left(\frac{1 + \cos\omega}{2}\right)^r \sum_{k=0}^{r-1} \binom{r + k - 1}{k} \left(\frac{1 - \cos\omega}{2}\right)^k. \tag{35}$$

The proof can be found in [20]. Other than the first few filters in the family, there are no simple formulae for finding the filter coefficients $c(n)$. Determining a spectral factor $\hat{c}(\omega)$ from the halfband filter $\hat{p}(\omega)$ is always possible, but not simple. The algorithms are given in [44].

EXAMPLE 10. Using the algorithm described we arrive at the Daubechies Db2 filter obtained in Example 9. For $r = 2$ the substitution $y = (1 - \cos\omega)/2$ in (35) yields the polynomial

$$\hat{p}(\omega) = \tilde{P}(y) = 2\,(1 - y)^2\,(1 + 2y).$$

From the quadratic equation

$$\frac{1}{2}\left(z + z^{-1}\right) = \cos\omega = 1 - 2\,y$$

we get roots $z_{1/2} = 2 \pm \sqrt{3}$ that correspond to the zero $y = -1/2$. The root with minor modulo $z = 2 - \sqrt{3}$ and the double root $z = -1$ represent three $(2r - 1 = 3$ for $r = 2)$ zeroes of the frequency response $C(z)$, so that

$$\begin{aligned} C(z) &= \alpha(1 + z^{-1})^2\,(1 - (2 - \sqrt{3})z^{-1}) \\ &= \frac{1}{4\sqrt{2}}\left((1 + \sqrt{3}) + (3 + \sqrt{3})z^{-1} + (3 - \sqrt{3})z^{-2} + (1 - \sqrt{3})z^{-3}\right), \end{aligned}$$

which is in accordance with (33). The coefficient α is determined by the condition

$$\hat{c}(0) = \sqrt{2}. \tag{36}$$

∎

By defining a lowpass filter, a highpass filter is defined as well by relations (25), looking like a mirror image of the lowpass filter – it has a multiple zero for $\omega = 0$ and a multiple value of one for $\omega = \pi$. Thus it is sufficient to deal only with the construction of the lowpass filter. The corresponding highpass filter on Figure 6.3 is $|\hat{d}(\omega)|^2 = |\hat{c}(\omega+\pi)|^2$. The sum of these two functions is constant, so there is no amplitude deformation. The flatness around $\omega = 0$ and $\omega = \pi$ yields great accuracy around these frequencies, while the accuracy is less around the middle. The greater the flatness, the closer the filter is to the ideal one (see Figure 6.2). The greater flatness means that the frequency response of the lowpass filter has a higher order zero at the point $\omega = \pi$. The filter bank is orthogonal and provides a perfect reconstruction.

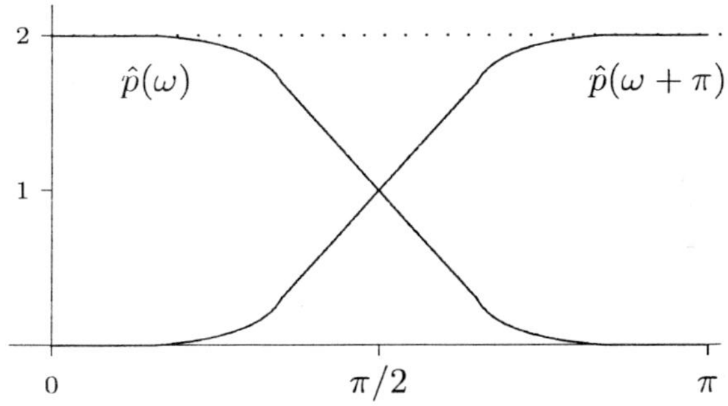

Figure 6.3: The maxflat filter bank

The filter orthogonality and flatness conditions can also be expressed through the filter coefficients $h(n) = c(n)/\sqrt{2}$ and the frequency response $\hat{h}(\omega) = \frac{1}{\sqrt{2}}\,\hat{c}(\omega)$,

CONDITION O (orthogonality)

$$\sum_{n=0}^{2r-1} h(n)\,h(n-2k) = \frac{1}{2}\,\delta(k), \qquad |\hat{h}(\omega)|^2 + |\hat{h}(\omega+\pi)|^2 = 1, \tag{37}$$

CONDITION A_r (approximation of order r)

$$\sum_{n=0}^{2r-1} (-1)^n\, n^k\, h(n) = 0, \qquad \hat{h}^{(k)}(\pi) = 0, \qquad k = 0, 1, \ldots, r-1. \tag{38}$$

Especially, from formulae (36) and (38) for $k = 0$ follows the property (4.20),

$$\sum_{\text{even } n} h(n) = \sum_{\text{odd } n} h(n) = \frac{1}{2}.$$

Daubechies filter family Dbr defines an important orthogonal wavelet family. They are described in §5.2.

6.5 Filter properties important for wavelets

From the previous analysis it is clear that there is a significant analogy between wavelets (continuous time) and filters (discrete time):

- multiresolution expressed by rescaling $t \to 2t$ has an analogy in the downsampling operator $(\downarrow 2)$;
- approximating by a scaling function corresponds to applying a lowpass filter to a signal;
- extracting details by wavelets corresponds to applying a highpass filter to a signal;
- an orthogonal wavelet basis matches an orthogonal filter matrix;
- a wavelet transform decomposes a function just like an analysis filter bank decomposes a signal;
- an inverse wavelet transform reconstructs a function just like a synthesis filter bank reconstructs a signal;
- a fast wavelet transform is calculated by multiplying with filter matrices.

Four filter properties play a central role in the wavelet theory.

Perfect reconstruction. The synthesis bank inverts the analysis bank with a delay of k,

$$F_0(z)\,\overline{H_0(z)} + F_1(z)\,\overline{H_1(z)} = 2\,z^{-k} \tag{39}$$

$$F_0(z)\,\overline{H_0(-z)} + F_1(z)\,\overline{H_1(-z)} = 0. \tag{40}$$

$H_0(z)$ is the lowpass and $H_1(z)$ is the highpass filter of the analysis bank, while $F_0(z)$ is the lowpass and $F_1(z)$ is the highpass filter of the synthesis bank,

$$H_i(z) = \sum_{k=-N_i}^{N_i} h_i(k)\,z^{-k}, \qquad F_i(z) = \sum_{k=-M_i}^{M_i} f_i(k)\,z^{-k}, \qquad i = 0, 1. \tag{41}$$

It follows from (39) that polynomials, with negative degrees of z as well, $\overline{H_0(z)}$ and $\overline{H_1(z)}$ have no common zeroes. Thus equation (40) is valid for every z only if the following conditions are met

$$F_1(z) = p(z)\,\overline{H_0(-z)} \quad \text{and} \quad F_0(z) = -p(z)\,\overline{H_1(-z)}. \tag{42}$$

Substituting relations (42) in (39) yields the equation

$$p(z)\left(\overline{H_0(-z)}\;\overline{H_1(z)} - \overline{H_0(z)}\;\overline{H_1(-z)}\right) = 2\,z^{-k},$$

only possible if $p(z)$ is a polynomial with the form $p(z) = c\,z^l$. When we put $c = 1$ and $l = -1$ in polynomial $p(z)$ in (42), we arrive at the relations

$$H_1(z) = z^{-1}\,\overline{F_0(-z)}, \qquad F_1(z) = z^{-1}\,\overline{H_0(-z)}, \tag{43}$$

which, if substituted in the equation arrived at by conjugating the condition (39), yield

$$H_0(z)\,\overline{F_0(z)} + H_0(-z)\,\overline{F_0(-z)} = 2\,z^k. \tag{44}$$

Identity (44) gives relation between lowpass filters, while highpass filters are then determined by formulae (43).

Orthogonality. This is a special case of the perfect reconstruction, when the analysis filter bank and the synthesis filter bank are the same

$$F_0(z) \equiv H_0(z), \qquad F_1(z) \equiv H_1(z). \tag{45}$$

Identity (44), considering (45), gives for $k = 0$ well known lowpass filter orthogonality condition (30),

$$|H_0(z)|^2 + |H_0(-z)|^2 = 2.$$

If we put in (42) that $p(z) = -z^{N-1}$ for even filter length N, to meet orthogonality condition (25), the highpass filter is equal to

$$H_1(z) = -z^{1-N}\,\overline{H_0}(-z).$$

The analysis bank is inverted by its transposed bank, i.e. the bank defined by the matrix transposed to the analysis bank matrix. The matching wavelets are orthogonal to all of their dilatations and translations.

Biorthogonality. If symmetry and perfect reconstruction are required, we have to use different analysis and synthesis filter banks. These filter banks are defined by biorthogonal filters that lead to symmetric biorthogonal wavelets (§5.3). Relation (44) for $k = 0$ is possible only if the double shift orthogonality condition for biorthogonal lowpass filters $H_0(z)$ and $F_0(z)$

$$\sum_n h_0(n)\, f_0(2m+n) = \delta(m), \qquad m \in \mathcal{Z},$$

is satisfied. The highpass filters $H_1(z)$ and $F_1(z)$ are completely determined by the lowpass filters from (43). Taking into consideration the presumed symmetry of the filters $h_i(k) = h_i(-k)$ and $f_i(k) = f_i(-k)$, as well as the assumption that the filter coefficients are real numbers, the highpass analysis filter response is arrived at

$$H_1(z) = z^{-1} \sum_k f_0(k)\,(-z)^k = \sum_j f_0(1-j)\,(-1)^{j+1}\, z^{-j},$$

and, similarly, the response of the highpass synthesis filter is equal to

$$F_1(z) = z^{-1} \sum_k h_0(k)\,(-z)^k = \sum_j h_0(1-j)\,(-1)^{j+1}\, z^{-j}.$$

By comparing the polynomials arrived at with (41) for $i = 1$, yields the relations of the filter coefficients (5.13) already derived for biorthogonal bases in §5.3,

$$h_1(n) = (-1)^{n+1}\, f_0(1-n), \qquad f_1(n) = (-1)^{n+1}\, h_0(1-n).$$

Maximum flatness. The frequency response of the filter has a zero of the order r in the point π,

$$\hat{h}_0^{(k)}(\pi) = 0, \qquad k = 0, 1, \ldots, r-1.$$

We name this property as A_r, the approximation of the order r, in wavelet theory. For a sufficiently smooth function this property produces the accuracy of the order r of the approximation determined by the scaling functions $\varphi(x-k)$. Also, it produces r vanishing moments of the wavelet $\psi(x)$. The decrease rate of the function decomposition coefficients by wavelets is of the order r for smooth functions, which ensures efficient representation. Finally, maximum flatness provides that the filter matrix eigenvalues satisfy necessary conditions for the existence and smoothness of the solution of the dilatation equation. More about it in §4.4.

Eigenvalues. These conditions have no importance for filters, but they are important for wavelet theory. They define other conditions per the filter matrix eigenvalues (besides those induced by the maxflat property), which guarantee the stability of the wavelet basis and determine wavelet smoothness. They also define the conditions under which the cascade algorithm (4.2) converges towards the scaling function $\varphi(x)$. As two time scales t and $2t$ represent a continuous form of downsampling $(\downarrow 2)$ $(x(n) \to x(2n))$, the cascade algorithm represents an iterative procedure defined by the matrix $M = (\downarrow 2)\, 2F$ from (4.12), where F is the filter matrix (2.31). The scaling function $\varphi(x)$ is determined by the infinite product of these matrices. More about this is given in §4.5.

An analysis of these properties from the aspect of filters is given in following examples.

EXAMPLE 11. Let the filter coefficients be $h(0) = 2/3$ and $h(1) = 1/3$. The cascade algorithm defined by this filter

$$\varphi^{(j+1)}(x) = \frac{4}{3}\varphi^{(j)}(2x) + \frac{2}{3}\varphi^{(j)}(2x-1), \qquad \varphi^{(0)}(x) = \begin{cases} 1, & x \in [0,1) \\ 0, & x \notin [0,1) \end{cases},$$

does *not converge*, because $\varphi^{(j)}(0) = (4/3)^j \to \infty,\ j \to \infty$. The point $z = -1$ (i.e. $\omega = \pi$) *is not a zero* of the frequency response

$$H(z) = \frac{2}{3} + \frac{1}{3}z^{-1}.$$

∎

EXAMPLE 12. (see Example 7) The box function is a *fixed point* of the cascade algorithm defined by the averaging filter $h(0) = 1/2$ and $h(1) = 1/2$, because in the first iteration it already yields a box function,

$$\varphi^{(1)}(x) = \varphi^{(0)}(2x) + \varphi^{(0)}(2x-1) = \varphi^{(0)}(x).$$

The frequency response of the averaging filter is

$$H(z) = \frac{1}{2} + \frac{1}{2}z^{-1},$$

and the point $z = -1$ (i.e. $\omega = \pi$) *is a zero* of the frequency response. The energy spectrum density

$$P(z) = 2\,H(z)\,H(z^{-1}) = \frac{1}{2} + \frac{1}{4}(z + z^{-1})$$

has no even degrees by z and z^{-1}, other than a constant, thus the filter is halfband. The box function *is orthogonal* to its translations. ∎

EXAMPLE 13. (see Example 8) The filter $h(0) = 1/4$, $h(1) = 1/2$ and $h(2) = 1/4$ defines a cascade algorithm

$$\varphi^{(i+1)}(x) = \frac{1}{2}\,\varphi^{(i)}(2x) + \varphi^{(i)}(2x-1) + \frac{1}{2}\,\varphi^{(i)}(2x-2),$$

which *converges* towards the roof function. The filter frequency response

$$H(z) = \frac{1}{4} + \frac{1}{2}\,z^{-1} + \frac{1}{4}\,z^{-2} = \frac{(1+z)\,(1+z^{-1})}{4}$$

has a double zero in the point $z = -1$. The filter is not halfband, because the product $H(z)\,H(z^{-1})$ contains even degrees of z and z^{-1}. The roof function is *not orthogonal* to its translations. ∎

These examples illustrate how properties of the filter frequency response, defined by the dilatation equation coefficients, determinate the convergence of the cascade algorithm and the properties of its solution, if it exists. We notice that the zero at the point $z = -1$ (i.e. $\omega = \pi$) of the filter frequency response is a necessary condition for the convergence of the cascade algorithm – if the response does not have a zero at the point $z = -1$, the cascade algorithm certainly does not converge (Example 11). Similarly, the orthogonality of the scaling function relative to the translations is impossible if a filter is not the halfband (Example 13).

7

Applications

Wavelets were independently developed in mathematics, quantum physics, electrical engineering and seismic geology. An exchange of ideas that have arisen in these areas brought about many new applications of wavelets in the last thirty years – for example, image processing, turbulence modeling, earthquake prediction, remote galaxy exploration or the discovery of similar behaviors in time series. In molecular spectroscopy they are used to remove noise from detected signals. They are also used in music to synthesis sound, in the examination of the fractal nature of objects, in fluid dynamics and, generally, in all areas where we are faced with complex structures with a multi-layer resolution.

Considering the fact that wavelet theory is a young scientific discipline, further expansion in the theory and application domain can be expected. To get an idea about variety of implementations only a few of the successful wavelet applications are listed below.

7.1 Signal and image processing

Wavelet approximation has primarily been developed in the last few decades for the processing of signals and images, as two-dimensional signals (§2.3). A signal may be, for example, temperature readouts. Speech represents changes in air pressure depending on time, while the complex graph of that function is the "adapted copy" of the voice. Signals also appear in telecommunication, images gained from satellites and medical recordings (echography, tomography and nuclear magnetic resonance). A discrete signal is a sequence of numbers obtained by measurement. One-dimensional signals are, usually, functions of time. Signal processing is everything that includes the analysis and interpretation of complex time series. Signals need to be analyzed correctly, coded efficiently, transmitted fast and then carefully reconstructed into fine oscillations or changes to the time function at the receiving end. Data compression appears as a consequence of the limited capacity of the transmission channels, as well as problems with data storage. Decoding, synthesis and reconstruction operations represent reverse operations to coding and quantization. With digital signals, based on a series of ones and zeroes traveling through a transmission channel, we need to reconstruct a signal or shape. Signal reconstruction has been compared to the restoration of old paintings. The artificial data and errors (noise) need to be removed, while certain characteristics of the signal that have disappeared by weakening or deterioration need to be amplified. Wavelets have a very important role in signal processing, no matter the source.

As an illustration, we shall list several examples of wavelet application.

Noise removal in data. The problem consists of discovering the real signal based on incomplete, indirect or noisy data. The cleaning of the signal consists in the removal of details with coefficients beneath a certain threshold, by replacing these coefficients with zeroes (see Example 5.2). Then, by an inverse wavelet transform we arrive at the cleared signal.

In this way, Ronald Coifman, with his colleagues at Yale University, by using a technique called adapted wavelet analysis, managed to clear the noise of an old record of Brahms' "Hungarian Dance" performed on the piano. The original radio record was completely unrecognizable.

Another example is from climatology. Based on temperature measurement results at various points on the northern hemisphere in the last two centuries, scientists are trying to examine the hypothesis of whether industry leads to global warming. The considerable natural temperature fluctuations represent a significant problem to be diagnosed and analyzed, and then removed from the registered signal to arrive at accurate data on the artificial warming of our planet as a consequence of human activity.

Seismology. One of the most striking features of seismic signals is their highly non-stationary character. This non-stationarity confounds traditional data analysis and processing tools, such as time-invariant filtering and Fourier transform techniques. The location and prediction of seismic activity is performed based on the wavelet coefficients of the seismic signal. The seismic signals warn of the possible ground vibrations of the Earth, an earthquake or explosion. These vibrations are waves, having their longitudinal and transversal components varying in various phases of the arrival of the earthquake. The seismograph registers both components and processes them using discrete time wavelet transform, enabling the location of seismic activity. The basis functions used for the analysis of the seismic signal need to be well defined in the time and frequency domain. Not only extreme seismic activities of the Earth are of interest. For example, seismic imagery of the earth's subsurface is critical to all aspects of the oil and gas exploration and production process – from the location of fields to their appraisal, development, and subsequent monitoring.

Industry. For centuries simple acoustic tests were used to examine the quality of an object's material, like striking a melon with a finger to test if it is ripe. Lately, modern acoustic analysis and signal processing equipment is used for quality control in highly automated factories. For example, computer hard drive manufacturers use Fourier transforms to check for irregularities in the sound during the high-speed rotation of the disks. This technique, however, cannot be applied to small non-periodical signals, which cannot be processed by the short time Fourier transform (STFT), either. During the eighties, simple experiments have shown that wavelet analysis can be successfully used there. Unlike STFT, wavelet analysis allows for an arbitrarily good frequency and time decomposition at high frequencies. This means that when, for example, a signal is made up of two bursts, they can be separated using wavelets if a high enough resolution is applied. Wavelet analysis also allows for a good and stable multiresolution representation and efficient numerical calculations, not possible with Fourier analysis.

These properties made Hisakazu Kikuchi and his colleagues from Niigata University use wavelet transform for the analysis of explosions within automobile engines. Explosions appear due to errors in ignition control when the engine is starting. They create shock waves that can sometimes even destroy the engine. Their discovery and analysis are important for the improvement of the ignition system. Data from the acoustic vibration analysis (the basic method used for studying explosions) contains false information, such as the noise created when setting mechanical parts in motion. The other method used for this purpose, statistical analysis, is also impractical, because the detonations are irregular. Wavelet transform of the sonic signal received at engine ignition, however, yields useful information. Kikuchi constructed a faster processor for the visual display of data arrived at through wavelet analysis.

In order to test the proposed solutions, two interesting experiments were performed. The first identified what was believed to be the characteristic components of the sound of an engine. Their synthesis yielded a sound very much alike to the real sound of the engine, confirming to the experts that the key components were identified successfully. The second experiment performed comparisons of the wavelet analysis capabilities and pressure sensor capabilities in discovering explosions. The pressure sensor, monitoring pressure data inside the engine cylinder, was the best instrument for finding explosions. However, its use in factories is very expensive because they need to be set up specially for an engine type, driving style (speed, acceleration), weather conditions, etc. It was shown that wavelet analysis is more accurate and efficient in finding explosions than pressure sensors.

A more banal, but no less important example is the use of wavelet analysis for the discovery of irregularities in the operation of a concrete mixer. Analyzers for the sound emitted during the rotation of the cylinder in the cement mixer can find irregularities. A further improvement of wavelet based techniques enables the construction of devices for reliable quality control.

Economy. The traditional approach to the application of spectral techniques to economic and financial data has focused on the discovery of complex, but stable frequency components. But, the problem is that in reality these components vary in strength, i.e. such signals are not stationer in time. Multiresolution approach in wavelet theory is also very important for modeling the dependence of physical relationships from the time scale, which is fundamental for analysis of economic and financial relationships.

For example, the difference in time horizon and its effects on bond holdings between short term money managers and those determining the investment portfolio for an insurance company. In fact, each agent operates on many scales simultaneously and some coordination between scales is needed. In [41] wavelets are used for the analysis of the cross covariances between scales on the money income relationship. The result obtained is that at shorter scales income causes money, that at intermediate scales money causes income, and that at longer scales there is a feedback mechanism.

Another example is analysis of the existing models. Most time series research assumes that the delays are fixed constants. Wavelet analysis shows that the "timing" of action by economic agents should not be a neglected, i.e. delays at certain scales are functions of time and are likely to be functions of the state space.

Wavelet analysis of various economic and financial signals enables deeper understanding of some well known results. It provides an explanation for a widespread phenomenon in economic and financial research in that while good fits can often be obtained, our ability to forecast is very poor. The scaling properties and the random nature of daily observations of the Standard and Poor market index are analyzed, which gives that the data are far more complex than previously assumed. Noise effects (details modeled by a few Dirac delta functions) that randomly come and go, are essential for the forecast. This result is very important in the analysis of the term structure of interest rates.

More examples and a comprehensive list of references can be found in [40].

Medicine. The electrocardiogram (ECG) is a measure of the electrical activity associated with the muscular contraction of the heart. Every phase of the contraction manifests itself as a wave. Analysis of the local morphology of the ECG signal and its time varying properties has produced a variety of clinical diagnostic tools for the discovery of irregularities in the working of the heart. Wavelet-based so called QRS detection methods are derived. More details and comprehensive list of references can bi found in [2].

Besides this medical aspect, the technical aspect of the use of wavelets as the ECG signal processing tool is very important. The compression of human ECG data enabling the establishment of ECG databases and transfer of ECG signals through telephone lines. A maximum efficiency with a minimum error is important for this. One of the criteria for the efficiency of compression is the ratio of the number of bits in the original data and the number of bits in the compressed data (PCD). Jie Chen and his colleagues from a Japanese university suggested an algorithm for the processing of ECGs based on wavelets where the PCD is between 5.5% and 13.3%. This algorithm is fast and simple, and further improvements are expected. Compression using that algorithm retains the important properties of the ECG signal that the conventional methods of transformation could not retain.

Wavelet based techniques have shown to be much better, especially when dealing with acoustic data. Echo-cardiography is a visual technique using ultrasound for the transfer of information. Ultrasound is an acoustic wave with a frequency above 20 kHz. For medical use ultrasound has a frequency of about 2MHz, providing finer precision with less penetration. Ultrasound penetrates non-homogenous environments and reflects off the edges of observed areas with various acoustic impedances. An echo-cardiograph notes the reflected sound. Using an echo-cardiograph we can observe the structure of the heart: the cardiac muscle, the inner and outer membrane, the ventricular septum. Using mathematical processing of the signal thus received, the frequency and acceleration of the operation of the walls of the cardiac chamber (ventricle and atrium) are calculated, providing clinical information on the working of the heart.

Image processing is performed on a numerical representation of the image, which is a matrix of the values of the function $f(x, y)$ in so-called "grey scale" at points of a sufficiently fine grid. When the image changes in time, a sequence of images needs to be analyzed, meaning that calculations need to be performed on an enormous amount of data. Thus it is necessary to note regularities and correlations that always exist between various parts of the numerical information representing an image.

As an example we can mention a study of remote galaxies. The origin and hierarchical organization of remote galaxies can be explained by wavelets. New telescopes provide a digitalized image of the Universe with an enormous amount of data. Wavelet analysis is used for the processing of these images. Object edges are determined, and the three-dimensional organization of the Cosmos is to be arrived at using them.

Fingerprint cataloguing is one example for the successful application of wavelets. The FBI archive contains approximately 200 million cards with fingerprints. Every fingerprint is digitalized to a resolution of 500 pixels per inch with a 256-level grey scale per pixel. Thus a single print takes up approximately 700,000 pixels, which amounts to approximately 0.6Mbytes of memory [26]. Thus, a large amount of data needs to be archived, but enabling fast searches. This means data compression, requiring a prior processing. The FBI has recently adopted a digital compression standard for fingerprint images based on wavelets. Compression is performed by the with a ratio of 20:1, and the differences between the original and compressed image can only be noticed by experts.

7.2 Numerical modeling

Last three decades wavelets have become a popular approximation tool in numerical modeling, although not so wide applied as in signal processing. There is a difference in the way wavelets are used in the processing of discrete and continuous quantities [45]. In the processing of discrete signals the sampling frequency, and thus the finest level of multiresolution analysis, are determined. We usually do not have information on the smoothness of the data. Thus asymptotic error estimates, usually based on the smoothness of the data and the fact that the resolution can be increased, are useless. In numerical analysis we are attempting to solve a mathematical problem formulated in terms of a function of a continuous variable. It is known that the solution has a certain smoothness, thus solution control is possible.

The following issues are of vital importance for a numerical modeling:

- Efficiency of the numerical solver, which means that the cost of the numerical solver is proportional to the degrees of freedom.

- Accuracy and reliability of the numerical approximation, which means that numerical solutions fit well relevant physical quantities.

- Data compression, which means efficient strategies to keep the discrete problems as small as possible without decreasing accuracy significantly.

The reason why the wavelets are not so wide applied is the higher complexity of the tool itself. To apply to a real life problem a wavelet basis has to be adapted to the considered problem and to the domain of interest, which is still much less advanced than for more conventional scheme. Main advantages of wavelet bases that recommend them for solving problems not accessible with conventional numerical methods, are:

- A set of basis functions can be improved in a systematic way. If one uses orthogonal or biorthogonal bases, the algorithms are simple and cheap, and usually more stable than in other methods.

- Different resolutions can be applied in different regions of a domain. The coupling between different resolution levels is easy. The only requirement is that a region of a higher resolution is contained in a region of a next lower resolution (one of the basic multiresolution properties).

Differential equations. Numerical models of complex processes and systems in science and technology are mainly based on differential equations, which are treated in various ways by wavelets [4, 9, 10, 14, 34]. Usually wavelets are used as basis functions for well known variational methods, Galerkin [36, 50], collocation [6, 48] or least square [18] methods. Because of the close similarities between the scaling function and the finite elements, both are functions with compact support, it seems natural to try wavelets where traditionally finite element methods are used, e.g. for solving boundary value problems. They are used to improve some finite element algorithms [11, 12]. It has been shown that for many problems a stiffness matrix condition number remains bounded as the dimension goes to infinity [17]. This is in contrast with the situation for regular finite elements, where the condition number tends to infinity. Wavelets are efficiently used for multilevel preconditioning.

Wavelets have also been used in the solution of evolution equations. Generally, in solving partial differential equations, time discretization is done using classical finite difference scheme of semi-implicit type. One obtains a set of ordinary or partial differential equations in space which are then solved at each time step by a weighted residuals method. The particular choice of wavelets as trial and test functions defines the different kinds of integration methods. Adaptivity can be used both in time and space.

Wavelets have shown good results in the numerical modeling of continuous values, especially those characterized by large gradients, typical for shock waves and turbulence problems.

Boundary layers. One class of problems, which requires a special numerical treatment, are so called singularly perturbed boundary problems. The reason is that a thin layer with a high solution gradient near the boundary or/and in the interior of the domain appears. Two approaches in construction of uniform convergent methods suitable for solving such problems, fitting operator and fitting mesh methods, can be combined by use of wavelets, as they can fit exponential nature of the solution and have multiresolution property [38, 39].

Turbulent flows. Analysis and simulation of turbulence belong to the most difficult problems in fluid dynamics. It is characterized by the occurrence of vortices or wiggles, whose size may vary over a large scale. A turbulent flow is modeled by the Navier-Stokes equation, in which the nonlinear advective term is larger by several orders of magnitude then the linear dissipative term (the large Raynolds number). A main problem for the numerical solution is the large range of scales present in the solution. A classical deterministic approach to compute a fully-developed turbulent flow separates by means of linear filtering between large-scale modes, assumed to be active, and small-scale modes, assumed to be passive. Statistical models are used to describe small-scale modes. To reconcile both points of view the space and scale decomposition based on wavelets is used. This new approach, called Coherent Vortex Simulation (CVS), is based on the nonlinear filtering defined in the wavelet space [23].

Operators. The wide classes of operators have sparse representations in wavelet bases thus permitting a number of fast algorithms for applying these operators to functions, solving integral equations, etc [8]. The operators that can be efficiently treated using representations in the wavelet bases include Calderon-Zigmud and pseudo-differential operators. Wavelets are used as approximate eigenfunctions of Calderon-Zigmud operators [3], which bring that transformation matrices related to such operators become almost diagonal. This enables compression of the operator by simply omitting small elements, similar to the case of image compression.

Bradley and Brislawn from the Los Alamos laboratory applied wavelets to calculating partial differential equations, but in a completely novel way. They did not use the wavelets to solve the equations, instead, they improved the functioning of the supercomputer while simulating the global climate and ocean model. They developed a vector method based on wavelets for multidimensional data sets of this model. In this case, the aim was to provide researchers with a rough, but readable interpretation of the supercomputer calculations.

More information about wavelets and their applications can also be found on the Internet, on one of the following web sites.

Web sites

1. bigwww.epfl.ch
2. www.amara.com/current/wavelet.html
3. www.amara.com/IEEEwave/IEEEwavelet.html
4. www.c3.lanl.gov/ brislawn/FBI/FBI.html
5. www.cs.dartmouth.edu/ sp/lift
6. www.mathworks.com/products/wavelet/
7. www-stat.stanford.edu/~wavelab/
8. www.users.rowan.edu/~polikar/WAVELETS/WTtutorial.html
9. www.wavelet.org
10. www.wolfram.com/products/applications/wavelet/

Bibliography

[1] Abramowitz M., Stegun I., *Handbook of Mathematical Functions*, Dover Publications, New York (1970)

[2] Addison P., *Wavelet transforms and the ECG: a review*, Physiological Measurement 26, 155-199 (2005)

[3] Alpert B, Beylkin G, Coifman R., Rokhlin V.D., *Wavelet-like basis for the fast solution of secon-kind integral equations*, SIAM J. Sci. Comput. 14, 159-184 (1993)

[4] Andersson U., Engquist B., Ledleft G., Runborg O., *A contribution to wavelet-based subgrid modeling*, Appl. Comp. Harmonic Analysis 7, 151-164 (1999)

[5] Barker V.A., *Some Computational Aspects of Wavelets*, Informatics and Mathematical Modelling, Technical University of Denmark (2001)

[6] Bertoluzza S., Naldi G., *A wavelet collocation method for the numerical solution of partiall differential equations*, Appl. Comp. Harmonic Analysis 3, 1-9 (1996)

[7] Beylkin G., *On wavelet-based algorithms for solving differential equations*, in Wavelets: Mathematics and applications (Eds. Benedetto and Frazier), CRC, Boca Raton, FL, 449–466 (1994)

[8] Beylkin G., Coifman R., Rokhlin V., *Fast wavelet transforms and numerical algorithms I*, Comm. Pure and Appl. Math. 44, 141–183 (1991)

[9] Beylkin G., Keiser J.M., *On the adaptive numerical solution of nonlinear partial differential equations in wavelet bases*, J. Comp. Physics 132, 233-259 (1997)

[10] Bihari B., Harten A., *Application of generalized wavelets: an adaptive multiresolution scheme*, J. Comp. Appl. Math. 61, 275-321 (1995)

[11] Canuto C., Tabacco A., Urban K., *The wavelet element method. Part I: construction and analysis.*, Appl. Comp. Harmonic Analysis 1, 1-52 (1999)

[12] Canuto C., Tabacco A., Urban K., *The wavelet element method. Part II: realization and additional features in 2D and 3D*, Appl. Comp. Harmonic Analysis 8, 123-165 (2000)

[13] Chui C., *Wavelets: A Mathematical Tool for Signal Analysis*, SIAM, Philadelphia (1997)

[14] Cohen A., Dahmen W., De Vore R., *Adaptive wavelet methods II - beyond the elliptic case*, Found. Comput. Math. 2, 203-245 (2002)

[15] Cooley J. W., Tukey J. W., *An algorithm for the machine calculation of complex Fourier series*, Math. Comput. 19, 297–301 (1965)

[16] Crochiere R.E., Webber S.A., Flanagan J.L., *Digital coding of speech in sub-bands*, Bell System Technical Journal 55, 1069–1085 (1976)

[17] Dahmen W.,Kunoth A., *Multilevel preconditioning*, Numer. Math. 63, 315-344 (1992)

[18] Dahmen W.,Kunoth A.,Schneider R., *Wavelet least square methods for boundary value problems*, SIAM J. Numer. Anal. 39, 1985-2013 (2002)

[19] Daubechies I., *Orthonormal bases of compactly supported wavelets*, Comm. Pure Appl. Math. 41, 909–996 (1988)

[20] Daubechies I., *Ten Lectures on Wavelets*, SIAM, Philadelphia (1992)

[21] Daubechies I., Sweldens W., *Factoring wavelet transforms into lifting steps*, Electronic, 1–27 (1997)

[22] Deslauriers G., Dubuc S., *Symmetric iterative interpolation processes*, Constr. Appr. 5, 49–68 (1989)

[23] Farge M.,Schneider K., Kevlahan N., *Non-Gaussianity and coherent vortex simulation of two-dimensional turbulence using an adaptive orthogonal wavelet basis*, Physics of Fluids 11, 2187-2201 (1999)

[24] Gabor D., *Theory of communication*, J. Inst. Electr. Engrg., London 93, 429–457 (1946)

[25] Goedecker S., *Wavelets and Their Applications for the Solution of Partial Differential Equations in Physics*, Presses Polytechniques et Universitaires Romandes, Lausanne (1998)

[26] Graps A., *An introduction to wavelets*, IEEE Comp. Science and Engineering 2 (1995)

[27] Grossmann A., Morlet J., *Decomposition of Hardy functions into square integrable wavelets of constant shape*, SIAM J. Math. 15, 723–736 (1984)

[28] Haar A., *Zur theorie der ortogonalen funktionen-systeme*, Math.Ann. 69, 331–371 (1910)

[29] Mallat S., *Multiresolution and Wavelets*, Ph.D. Thesis, University of Pennsylvania, Philadelphia (1988)

[30] Mallat S., *A theory of multiresolution signal decomposition: The wavelet representation*, IEEE Trans. Pattern Anal. and Machine Intel. 11, 674-693 (1989)

[31] Meyer Y., *Ondelettes*, Hermann, New York (1990)

[32] Meyer Y., *Wavelets - Algorithms & Applications*, SIAM, Philadelphia (1993)

[33] Misiti M., Misiti Y., Oppenheim G., Poggi J-M., *Wavelet Toolbox (for use with MATLAB*, The MathWorks, Inc.,Natick, Mass. (1996)

[34] Piquemal A. S., Liandrat J., *A new wavelet preconditioner for finite difference operators*, Advances in Comput. Math. 22, 125-163 (2005)

[35] Qian S., Weiss J., *Wavelets and the numerical solution of partial differential equations*, J.Comp. Physics 106, 155–175 (1993)

[36] Suen J., Nayak R., Armstrong R., Brown R., *A wavelet-Galerkin method for simulating the Doi model with orientation-dependent rotational diffusivity* J. Non-Newtonian Fluid Mech. 114, 197-228 (2003)

[37] Radunović D., *Numerical methods*, (serb.) Akademska misao, Beograd (2004)

[38] Radunović D., *Spline-wavelet solution of singularly perturbed boundary problems*, Mat. vesnik 59, 31-46 (2007)

[39] Radunović D., *Multiresolution exponential B-splines and singularly perturbed boundary problem*, Numer. Algorithms 47, 191-210 (2008)

[40] Ramsey J.B., *Wavelets in economics and finanse: past and future*, Economic research reports, RR2002-02, 1-70 (2002)

[41] Ramsey J.B., Lampart C.,*The decomposition of economic relationships by time scale using wavelets: money and income*, Macroeconomic Dynamics 2, 49-71 (1998)

[42] Sabet K., Katehi L., *An integral formulation of two- and three-dimensional dielectric structures using orthonormal multiresolution expansions*, Int. J. Numerical Modelling 11, 3-19 (1998)

[43] Strang G., *Wavelets and dilation equations: a brief introduction*, SIAM Rew. 31, 614–627 (1989)

[44] Strang G., Nguyen T., *Wavelets and Filter Banks*, Wellesley-Cambridge Press (1996)

[45] Sweldens W., *The Construction and Application of Wavelets in Numerical Analysis*, Ph.D. Thesis, Leuven (1995)

[46] Sweldens W., *The lifting scheme: A custom-design construction of biorthogonal wavelets*, Appl. Comput. Harmon. Anal. 3, 186–200 (1996)

[47] Sweldens W., Piessens R., *Asymptotic error expansion of wavelet approximations of smooth functions II*, Numer. Math. 68, 377–401 (1994)

[48] Vasilyev O., Bowman C., *Second-generation wavelet collocation method for the solution of PDE*, J. Comp. Physics 165, 660-693 (2000)

[49] Vetterli M., Kovačević J., *Wavelets and Subband Coding*, Prentice Hall, Englewood Cliffs, New Jersey (1995)

[50] Yuesheng X., Qingsong Z., *Adaptive wavelet methods for elliptic operator equations with nonlinear terms* Advanced in Computational Mathematics 19, 99-146 (2003)

Index

Izdavač

AKADEMSKA MISAO

Bul. kralja Aleksandra 73, Beograd
tel./fax: (+381 11) 3218 354

office@akademska-misao.rs

www.akademska-misao.rs

CIP - Каталогизација у публикацији
Народна библиотека Србије, Београд

CIP

517.443

RADUNOVIĆ, Desanka P., 1950 -
Wavelets : from math to practice / Desanka P. Radunović. - Beograd : Akademska misao ; Berlin ; Heidelberg : Springer-Verlag, 2009 (Beograd : Planeta print). - X, 163 str. : graf. prikazi ; 24 cm

Tiraž 500. - Bibliografija : str. 157 - 160. - Registar

ISBN 978-86-7466-345-5 (AM)
ISBN 978-3-642-00613-5 (SV)

a) Вејвлетс (математика)
COBISS.SR - ID 156018700